Pieces of the Puzzle

Part 1

How My Dad Beat Cancer

Lijlanie

Pieces of the Puzzle

Part 1

How My Dad Beat Cancer

Printed in New-Zealand
by
Customer Focused Solutions Ltd.
Level 1, 36 Paramount Drive, Henderson 0610

Editing by Colleen Kaluza of WordWyze Publishing Ltd. WordWyze.nz
Front Cover by Marlise Meyer Photography
Cover design by Michelle Hugo michelle.hugo93@gmail.com
ISBN 978-1-99-118720-8 (Afrikaans paperback)
ISBN 978-1-99-118721-5 (Afrikaans E-book)
ISBN 978-1-99-118722-2 (English paperback)
ISBN 978-1-99-118723-9 (English E-book)

www.lijbeatslympho.com

Scripture quotations are taken from:

To my dad

*"The truth is like a lion.
You don't have to defend it.
Let it loose.
It will defend itself."*

~ Saint Augustine ~

Acknowledgments

Firstly, and above all, I offer thanks to our Triune-God. He is the Origin, the Inspiration, the Instructor of this book.

The past six years would have been insurmountable without Aaron and Hur.

A special thank you to my husband and children, who were simply there during the darkest period of my life; to my family of faith, childhood-friends and all the other friends who came along later and carried me through this time.

Oom Louw, my dad's childhood-friend and soul mate, thank you for keeping me accountable chapter by chapter. Your support gave me the courage to complete this task.

Marlise Meyer and her family is responsible for the cover photo.

Marcel Versveld, thank you for your advice and help.

To all of you who urged me to translate this book into English: if it wasn't for you, the English translation may not have materialised.

Thank you to my beta readers, Pastor Jan Rodgers and Liezl Wienand. Your feedback and encouragement are much appreciated.

Colleen Kaluza, editor of the English version of this book, I loved working with you. Thank you.

Finally, in the footsteps of our Heavenly Father – my parents, Chris and Petro Viljoen: after God, you are the foundation upon whom my values and humanity is founded. I will honour you always. *Mamma* encouraged me to write a book since I was a little girl and *Pappa* taught me the perseverance to complete this task.

May 'Pieces of the Puzzle' honour your legacy.

Foreword

In sharing her story, Lij manages to lay bare close-to-home, sensitive and deeply rooted emotion with conscious and perceptive observation and brutally honest insight.

She does not try to escape when the intimate detail of her journey gets too close for comfort. Instead, she welcomes the reader as a fellow pilgrim with her courageous, introspective testimony.

By capturing traces of childlike emotions and precious pieces of her childhood, she creates a theatrical ambience where spiritual warfare reaches climactic intensity. It is this unrestrained testimony that weaves integrity, remarkable talent, and the gift of grace into one to provide truthful nourishment for the weary fellow pilgrim.

The complexity of the circumstances makes the human narrative distinctive. The exceptional bond between father and daughter who brave the purifying crucible together is poignant and priceless. Especially to me, who shared Chris' friendship from an early age and who not only experienced, but also understood his notable, solid, and encompassing love for his family.

Despite her own incompleteness and human weakness, Lij testifies that the Almighty is in charge and bestows both grace and mercy to all who cling to Him.

Therefore, one can ask with believing boldness: "Lord, please let Your rain fall where my sheep are grazing..."

Well done. Receive my heartfelt thanks.

With best wishes,
Louw du Toit
Nylsoog
October 2022

Endorsements

"A beautiful, yet raw journey of valley and mountain top experiences, all laced with hope. I have been personally encouraged and inspired by Lijlanie's strong faith and her wisdom in having the right people on the journey to pick you up when you feel knocked down. I thank you, Lijlanie, for allowing us, the readers, to learn and glean from you and your dear family and for sharing the beauty of a miracle!"

~ Ps. Jan Rodgers ~
Activate Church, Hamilton, NZ

"I thoroughly enjoyed this book. It really touched my heart as Lijlanie brought the pieces of the puzzle together. Her story inspired me to choose my Aaron and Hur and to fight the good fight of faith, to trust in what God has said to me, even when in the natural things may seem different. A wonderful reminder that we serve a faithful God!"

~ Liezl Wienand ~

Ode to Aaron

My better half.
My voice of reason when I don't want to reason anymore.
My shoulder to cry on; my strength when I'm weak.
My look-alike. *My ander ek…*

How would I ever manage this journey without you?

"And God saw that it wasn't good that she was alone."
So He created one like me – my look-alike – my equal.
One whose spirit resonates with mine.
One who wages war like I do,
Believes like I believe, hopes like I hope…and mourns like I mourn.

Aaron – my voice when I'm too weak to breathe a whisper.
Aaron – the one keeping my arms raised.
Aaron – the one who throws her staff on the ground
to become a serpent,
Devouring all the serpents threatening me.
Aaron – blood of my blood.
Aaron – the one with whom I can be without filters.

And even though oceans separate us,
You continue to remain but a nano-second away,
And our thoughts jump - as in quantum physics -
Between each other with zero passing of time...

Contents

Chapter 1
The Gate

Friday evening, 12 April 2019. Family Movie Night! A beep sound indicates that the oven has reached the desired temperature. It brings my attention back to the present where I am standing in my brand-new kitchen in our brand-new house. For a moment, I was back at my house on the riverbanks ...

"*Ag nee, Mamma*, I am going to call Dad now to ask what time he will be home." My almost-teenage daughter's voice sounds rather impatient. The little missy grabs her mobile, and I can vaguely hear the ringtone. I stare out the window, once again amazed at the exquisite shades of pink and purple the Twilight Artist conjures up each evening. This is my favourite time of day: the peaceful hush when the birds quiet down as they return to their nests, just before the crickets awaken.

"Dad, what time will you be home?" Niki's voice interrupts my thoughts. "What?" Her cheeky voice sounds rather indignant.

The mother in me is quick to correct: "*Nie* 'what' *nie*, Niki! 'I beg your pardon, Dad.'" Although Afrikaans and English mingle leisurely in our Afrikaner-Kiwi-home, the language purist and home-schooling mum in me endeavours to ensure that each language is spoken correctly (and with the required respect!)

With rolling eyes, she corrects herself (well, in her own words, of course!): "Sorry, Dad, I didn't hear what you said? ... *Vyf minute?* ... Okay, Mum's putting the pizzas in the oven now ... Okay, see you in four minutes ... Bye."

"I thought Dad said five minutes?" I ask with a twinkle in my eye.

"*Ja*, but we have been chatting for at least a minute," Niki argues.

"Well, all right then! Please grab us some plates. We only have three minutes left now before he arrives," I tease, winking lovingly at her.

"In a minute, *Ma*, I just want to log into Showmax so long," she replies.

I glance through the glass oven door and shake my head while grabbing the plates myself. My darling child - always doing things in her own order. Forever figuring out a better way to do things than the suggested one. This is indeed a very useful personality trait – as long as it doesn't emanate from a child that one is trying to raise!

Out of the corner of my eye I spot hubby's car as it turns into our driveway. "*Mamma*, Daddy's home!" A little whirlwind rushes past me as Niki eagerly runs to the garage to share the daily news with her dad before he can even get out of the car.

A few minutes later, everyone has greeted one another, and we gather in the TV lounge. With a pizza on her lap and Showmax on pause, Niki asks: "So, Dad, what was your high?" Clearly, she cannot wait to share her own highlight of the week. However, her self-imposed order determines that the eldest has to share his highs and lows first.

"I suggest we first pray over our food, before it cools down too much and then we can start sharing highs and lows," I gently interrupt. We hold hands as we thank the Lord for the week gone by and for every blessing He bestowed upon our family. We pray that He will bless those less fortunate than us - especially the little children who may not be so lucky as to enjoy pizzas right now and who may have to go to bed hungry tonight.

My thoughts go back to the street kids outside the disco, Masquerades, where we used to sell pizzas from a mobile pizza oven when we were students, trying to earn some pocket money. The street kids always came begging for some money. Since we knew from experience that they would only use it to go and buy glue to get high on the toxic fumes, we thought it wiser to give them a free pizza. Sacrificing the few rands we would have made from selling it instead, didn't make us any poorer. At least it kept a few street kids from sniffing glue and quieted the hunger growls for a little while.

"*Ma*! We're waiting for your low?" Niki's words pull me back from my walk down memory lane. Each of us share our highs and lows of the week, while relishing our delicious homemade pizzas. A few minutes later, the well-known tune of *'Getroud met Rugby'* announces the intro to our favourite Showmax-series. Before we know it, we once again become intrigued in the drama-filled lives of the Bekker family and co.

5:23 pm: My phone rings. I am blissfully unaware of the incoming call since it is lying on the kitchen counter and on silent mode.

Moments later, hubby's phone starts to vibrate. I notice him glancing at the number, but I don't pay much attention. As a rule, we do not answer phone calls during our Friday Family Movie Nights. Quite unexpectedly, he passes me his phone: "It's your sister."

Fear clenches its ice-cold fist around my heart. Lin will only call me on hubby's phone if she cannot get hold of me on mine, and if it is very urgent. Besides, phone calls from South Africa at half past five New Zealand time in the evening on a weekday often send chills down one's spine. Too often, the African night has been cruel again ... *Lin is supposed to be at school, doing the early morning run-around before she starts teaching. Why is she calling me this time on a Friday morning?*

"Hello, Sis!" My cheerful voice is tainted with caution.

"Sis, don't be frightened," Lindi replies with a shaky voice. *Is she trying to soothe my nerves or her own?*

"A steel gate came crashing down on *Pappa*'s head," comes the shocking news.

"*Nee!*" I yell. *Ag, Here! My pa!*

"*Pappa*... is okay. He is at the Emergency Unit at present. The doctor is busy stitching him up. He is just very confused." Her voice breaks as we both start to cry.

"What happened, Sis?" I ask as countless thoughts and scenario's run through my mind. *Ag, Here! My pá!*

"*Pappa* usually opens the shop at half past six in the morning. I was busy packing the kids' lunchboxes when a stranger called me from *Pappa*'s phone. He just said a steel gate crashed onto *Pappa*'s head and that I had to hurry. Sis, when I arrived, there was so much blood! A few men helped get *Pappa* into my car and urged me to get to the hospital as fast as possible. I am now on my way back to the shop to try and find out exactly what happened. I just wanted to call you first." There is a moment of silence while Lin is breathing heavily.

So many gaps in the story. *Where's my mum? How on earth did the gate crash onto Pappa's head? Why did the guy… who is the guy? Why did he phone my sister and not my mum? Who is with my dad at the moment?*

"Where is *Ma*?" My voice quivers in my own ears.

"She is with *Pa* now. They couldn't reach her, so *Pappa* told them to phone me. Fortunately, he did not lose consciousness and could tell them who to call. I phoned *Mamma* from the EU."

Thank goodness Lin did the sensible thing and didn't waste any time trying to find *Mamma* first. Although they live right next to each other, *Mamma* would have probably been in the shower at that time of the morning. *Who is this man?*

"Sis, *Pappa* is very confused. I think he has a cracked skull because the cut is very long and deep." Her voice starts to quiver again.

The spiritual warrior inside of me takes over. "No!" I exclaim firmly. "He does not have a cracked skull! He is way too *hardekop*! We are proclaiming life over this situation. *Pappa* won't even have a headache!"

Immediately the warrior inside Lin rises up as well. We agree to speak life into the situation. "I will phone you the moment I know more," she promises.

As I end the call, my knees give way, and I sit down on the edge of my bed. I become still before God. "Talk to me, Lord?" The little girl in me wants to break down into hysterical sobs, but the eldest daughter in me needs to listen and hear what the God of the angel armies has to say.

I wait. Silence. I hold my breath. As if from a distance I hear the Voice from Eternity, who was and is and already knows the future, whisper to me: "I, the Lord your God, command you: 'Be strong and courageous. Do not be afraid or discouraged, for the LORD your God is with you wherever you go.'[1]

Peace beyond all human comprehension enfolds me. All of a sudden, I know without a doubt that my dad will be okay!

"Give your dad the following message from Me, so that he can be certain of this: I am with him. He will be okay." This is one of the few instances where I perceive God's voice as audible. It is as though my spirit is connecting with the spiritual realm where God's Voice can be heard loud and clear. "Remember, Lijlanie, My Spirit intercedes for you with mortifying groans. Not only My Spirit, but Jesus Christ Himself – the same Jesus who was crucified, but more than that, the One whom I resurrected from death to life, is sitting at the right hand of God. He is also interceding for you, according to the will of the Father."[2]

Without hesitating, I text Lin: "Sis, *Pappa* will be okay!"

*

Merensky High School, Tzaneen. I am four years old, and we have just returned from the *kafee*. *Pappa* bought a litre of Coke. I am very big and strong and can do almost anything all by myself. "*Pappa*, may I carry the Coke?" I beg urgently. *Pappa* hands me the one-litre glass bottle.

"Walk slowly. Be careful! If you drop this bottle, it will break, and you are *kaalvoet*," warns *Pappa* with a strict voice. *Mamma* is carrying the baby and lets me be. Full of confidence, I pitter-patter towards the concrete wastewater ditch. It was only yesterday when Carla and I practised to see who could take the longest leap over the ditch. Wait, let me see how far I can jump today!

With exhilaration, I approach the ditch and give one giant leap… My foot lands just short of the other side, and before I know it, the Coke bottle shatters into pieces at my feet. Coke splatters cover my legs and pants. My eyes grow wide, and my face turns pale. I try to wipe the sticky liquid from my legs with one hand and turn around to apologise to *Pappa.* He is going to be so angry with me!

As I opened my mouth to beg for his forgiveness, I feel a sudden, sharp pain in my foot. I look down to find a pool of blood gathering underneath my tiny little foot. Blinded by tears, I feel how my chest starts to explode as giant sobs burst forth - the fear of the upcoming reprimand far greater than the pain in my foot. Before I know it, I am in *Pappa*'s arms. He holds me tightly to his chest and whispers soothing words of comfort. *No reprimand?* I sob even louder. *Pappa*'s shirt grows wet underneath my cheek. His arms around me always makes me feel so warm and safe. My *Abba-Pappa*.

*

12 April 2019 – Circle of Trust – WhatsApp® group chat:

> 5:35 pm: *Mamma*: "At the hospital. Gate crashed down onto *Pappa*'s head outside the shop. Cannot remember anything. About a five-inch cut. They need to do a scan. He's asking the same questions over and over again. Lin went back to the shop to get more info. Talk later."
>
> 5:36 pm: Me: "Okay, *Mamma*. I am praying continuously. *Pappa* will be okay. Please keep me up to date. Let me know when I can give you a call."
>
> 6:18 pm: Other Sis: "Where is the Emergency Unit?"
>
> 6:23 pm: Me: "Whoever gets my message first, please send me a voice note and explain to me exactly what happened? Lin, the line was very unclear when I just spoke to you. Please give me an update on *Pappa*'s current condition. I need to know what's going on. *Dankie, julle.* Send *Pappa* all my love and shower him with kisses and hugs from me. I wish I was there …"

I scroll through the files in my brain. My brain's filing cabinet is such a mess right now! Where on earth are the words of the Voice written down again? Blank pages taunt me. *When all else fails, ask Doctor Google*. With clumsy fingers I try to type in the words: "I, the Lord your God, command you". I press the search key.

Of course! How could I forget? Joshua 1 verse 9! The rest of God's promise and the blessed assurance that not only the Holy Spirit but Jesus also are interceding for us according to the will of the Father is somewhere in Romans 8. I search for the relevant verses and phone *Mamma*. Voice mail!

6:25 pm, I hear a notification coming through. It is a voice message from Lin. They are still at the EU, where the emergency staff are stitching up my dad. *Pappa* is extremely confused and cannot remember anything about the incident. Once they finish the stitches, they will do a brain scan. The cell phone signal in the EU is very weak and I have to wait.

A little while later, I receive a phone call from Lin. She sounds calmer now, has better cell phone signal and can fill in the gaps for me.

"Sis, please explain to me in detail what happened. I have so many questions," I ask with a soft voice.

"Sis, during my quiet time with the Lord this morning, I battled to find the right words to pray. The only sentence I was able to put together was that I am cancelling the devil's plans against us in Jesus' Name. It was as if these words were the only ones running through my mind.

"Later on, while I was busy in the kitchen, I heard *Pappa* leaving for work. It was just after six, and I thought to myself that it was a bit early for him to leave. Throughout our early morning preparations, my son and I were listening to "I raise a hallelujah"[3] and were singing along. Fortunately, I finished all my early-morning duties. I was just about to wake up Edri with a cup of coffee when my phone rang. According to the caller id, it was a phone call from *Pappa.* But when I answered, it ... it wasn't *Pa* on the other end! I instantly knew that something was wrong."

I swallow hard on a huge lump in my throat, my chest tightens, and I battle to breathe. In my mind, I am reliving the traumatic moments with her.

"The voice of a young male urged me: '*Hoor hier*, you better rush to your dad's workshop.' Apparently, they couldn't get hold of *Mamma*. For some reason, her phone didn't ring that morning. I jumped into my car and left in a hurry. While I was driving, I felt the urge to play "I raise a hallelujah". You are familiar with the story behind the song, hey?"

"*Ja, ja*, the story of little Jackson who wasn't supposed to survive the night," I answered quickly, anxious to hear the rest.

"Well, I put it on continuous play. As I turned the corner close to the shop, I saw *Pappa*'s *bakkie*; I saw the gate lying on the ground and about seven young men crowding *Pa's bakkie.* As soon as I pulled into the driveway, one of the guys rushed over and urged: "'*Hoor hier*, you need to get your dad to the hospital. That gate came crushing down onto his head.'

"I jumped out of my car and ran towards the *bakkie. Pappa* was busy getting out of the *bakkie* with a cloth wrapped around his head. I asked whether he was okay, and he answered: '*Ja*.' The roof of my little car is quite low. After *Pappa*'s back surgery, his back is not very supple, so he usually battles to get into my car. However, this morning, Sis, he managed to get into my car with one swift move! There was so much blood! I grabbed a shirt from the backseat to prevent blood getting everywhere. But with that quirky move, the blood didn't even touch the headliner."

I giggle, despite all my tears and tight chest. Lin takes a moment to catch her breath. "Uhm," she mutters, trying to order her thoughts.

"The hospital underwent a whole make-over, so I didn't know where the new Emergency Unit was. Initially, I parked in the wrong parking bay. I was directed towards a different entrance. Fortunately, I remembered to grab *Pappa*'s cell phone and wallet, so I had all the information required to complete the admission forms at hand. My hands started shaking uncontrollably as I tried to complete the forms." *Your voice is shaking as well.*

I look down at my own hands, and realise they are shaking. I remind myself to breathe.

"I realised I needed to return to the shop to unlock and let the workers in," Lin continued.

My poor sister, you thought about everything, while all you wanted, was to hold Pappa's hand.

"This time, *Mamma*'s phone did ring when I phoned her from the EU. When she arrived there, I was able to return to the shop. On my way there, I phoned you. At the shop, Mary, Bettie and Shephard waited anxiously. After I explained to them what happened, I said: 'I think we need to pray.' We held hands and prayed together. The poor workers were very emotional."

"*Siestog*, they are very fond of *Pappa*. He never lets them start the day without reading from the Bible and praying over them," I add.

"The guy who owns the business across the road from our shop offered to keep *Pappa's bakkie* at his premises until we could collect it. After the workers and I prayed for *Pa,* I crossed the street to ask him what exactly happened this morning. According to him, he was at work earlier than usual. While he was reversing his *bakkie* up the driveway leading to his shop, he saw how *Pappa* was opening the gate, and how it pulled out of its hinges and fell right onto his head. Out of the blue, seven big young men appeared and lifted the gate off *Pa.* I guess we'll never know how many of them were, in fact, angels."

The gate motors in South Africa all too often 'go on midnight walks and get lost', never to return. That is why *Pappa* had to push the otherwise motorised gate open instead of pressing the remote.

"The barbed wire was tangled around *Pappa*'s head," Lin continues. *Ja, gates alone won't keep the intruders out.* "The men had to untangle the barbed wire carefully."

"*Jinne*, Sis, he could have lost an eye!" I exclaim, shocked.

"*Ja*, it is a miracle that his face wasn't mutilated. I thanked the men who were still present and started to cry. Then I called the school principal. By the grace of God, he didn't expect me to return to work today. It was then that I realised I hadn't phoned other sis yet. *Siestog*, she was distraught."

Lin becomes quiet. We are fraught with emotion – full of thoughts of what could have been and possibly still might be. Yet, we are overflowing with thankfulness that our dad is still alive.

"I need to get back to the hospital. 'Will keep you posted."

I press the call button next to *Mamma*'s name. Nothing. Of course! There is no signal at the EU. Frustration builds up in me. A prayer via a WhatsApp® voice note is the best I can do for now.

6:36 pm: Me: "'(I, the Lord your God, command you): 'Be strong and courageous. Do not be afraid; do not be discouraged, for the LORD your God will be with you wherever you go.'[4]

"*Pappa*, this is the Scripture the Lord gave me when I prayed for you just now. It was a massive shock to hear the news about what happened to you this morning. Lin phoned me and told me about the accident. I am praying for you constantly and know that you will be okay. An immense sense of peace overwhelmed me whilst praying for you. I just knew that you will be all right. I thank God that He is with you every step of the way. I also thank Him for urging Lin to deny the devil's plans against you in Jesus' Name, this morning.

"*Pappa,* I am so sorry for all the pain and discomfort you have to endure. I am sorry about what happened but thank the Lord that the devil cannot touch your life! *Ek is baie lief vir Pappa*. I know you are going to be okay. I am so sorry for the huge headache you may have, but in Jesus' Name I cancel that headache. You will be just fine. Jesus covers you in His blood and He carries you in His arms.

"I desperately want to be there! I am so sorry that I'm not! Thankfully, my mum and sisters are there. I know they are busy showering you with love and affection. I love you and I am praying continuously. Even when I go to sleep, you will still be in my prayers in my dreams.

"Jesus Himself is sitting at the right hand of God. He is interceding on our behalf. The Holy Spirit joins Him in intercession with mortifying groans. He knows our every need and desire. I call upon God as *Yehova Rapha* – the Lord our Healer. I thank God for His hand upon your head and your whole body and that He will heal you in Jesus' Name. I thank the Lord that you will have no side-effects, suffer no consequences as a result of this accident, and that you will be healed in the sweet, sweet Name of Jesus, our Lord and Saviour.

"I thank the Lord that I will go and fetch my dad from the airport very soon for that long-awaited visit. *Ek is baie lief vir Pappa*. In my spirit and in my heart, I am right there with you.

"I am with you in spirit, *Mamma* and with you, my dear sisters. Just close your eyes for a minute, and you will sense my presence within your spirits. *Ek is by julle*. I am with you. I love you all so much. As I am thinking of you and praying for you, in my mind's eye I can see you all covered by the blood of Jesus. Hence, you are invisible to the enemy, for you are hidden within the folds of His garment …

"*Pappa*, behave and take it easy. Fortunately, we know you are way too *hardekop* to suffer any permanent damage to your skull or brain. (I am saying this with all due respect and lots of love). *Ek het Pappa baie lief*. Jesus is with you. His peace is with all of you. I can visualise you all, standing around *Pappa*'s bed – a halo of peace surrounding you. The plans of the enemy will fail. He comes to steal and destroy…" my voice starts to break. "But Jesus came," I proclaim as my voice grows stronger and louder, "so that we can have life in abundance![5] Thank you, Jesus!"

In reply, Lin sends a link to the Bethel video, "I raise a hallelujah". I play it at maximum volume and sing along even louder.

*

We spent a lot of my childhood April school holidays in Durban. Most of the time, we stayed at the Four Seasons Hotel. I felt very important when the men in their burgundy jackets and waistcoats welcomed us at the entrance and took our bags. One of them was very tall, with two huge holes in his ears. Another one was very short, with a funny-looking complexion. *Mamma* always reprimanded me in a lowered, strict voice if I stared for too long at these men.

Whenever I saw the *ricksha*[6] boys in the street, I nervously stood closer to *Pappa* and grabbed his hand. I stared wide-eyed at the impressive *gedoente* on their heads, unsure whether I should laugh or be afraid.

April, 1979: As soon as the hotel room door closes behind us, I start showering *Pappa* and *Mamma* with questions.

"*Pappa*, when can we go to the beach? *Mamma*, where is my costume[7]? Can we go buy *spookasem* and ice cream? *Pappa*, can we really ride on the *riksha*? Won't we fall off? Remember, *Pappa,* you promised that I can go ride bumper cars and *spooktrein* with you. I am a big girl now!" With pride I hold up my right hand with five fingers stretched out wide.

It is already late in the afternoon and overcast, so *Pappa* suggests we take on the bumper cars and the *spooktrein.* There are more than enough fun rides for little kids. However, the *spooktrein* and bumper cars would leave such a lasting impression that I would think up excuses to avoid them throughout my school career.

Let's start with the *spooktrein*. My bravery disappears like early morning fog in the heat of the sun as soon as the first monster appears in front of my face. I scream and try to hide my face under *Mamma*'s arm. My eyes are tightly shut as I pray that God would quickly end this ride. Suddenly, I feel soft, sticky, long strands tickling my face. Hysterically, I start to cry. *Mamma* tries to make light of it and encourages me to open my eyes. I protest vehemently. She assures me the bad stuff is over. Cautiously, I peep through squinted eyes and realise I am trapped within a rotating kaleidoscope. I feel nauseous. Once again, I shut my eyes and only open them again when I feel the warm Durban air on my face and recognise the salty smell of the ocean air.

Ouma awaits me with open arms. Highly distressed I give her my account of the ride: "*Ouma*, I promise this *spooktrein* will never see me again – ever!"

Her soft lap is comforting. She assures me that she doesn't like the stupid *spooktrein* either and that it is an evil human creation. I am in complete agreement and together we plan how we are going to exterminate this monster from hell.

"Would you rather go ride bumper cars with me?" asks *Pappa*.

"*Ja! Ja! Asseblief, Pappa*!" I jump up and down with excitement.

However, after today, the bumper cars – just like the *spooktrein* – were going to wait until my early twenties to receive a visit from me again! One moment, I was laughing away with *Pappa* and the next moment, I only saw flashes of light. In the seventies, safety belts for kids were not high on the list of priorities. The first big bump sends me flying forward, and my head collides with the dashboard. Needless to say, our ride ends before the timer runs out. (Our visit to the amusement park as well.)

Back in the hotel room, I have my first opportunity to look into the mirror. A huge blue bump on my forehead stares back at me. I burst into tears: "Now I look ugly!" I cry. *Pappa, Mamma* and *Oupa* try hard not to laugh, but *Ouma* had just about enough of this amusement trip which was borderline-child abuse, according to her. "*Pappa* is going to buy himself a shotgun to shoot the *spooktrein* and the bumper cars to pieces!" she promises. I share this promise of revenge to everyone who looks the least bit interested in my 'ugly' forehead. My *Gibbor-Pappa*!

Early the next morning, I remind *Pappa* of our revenge quest: "Are you going to buy the gun today, *Pappa*?"

"No, today we are going to go to the beach," *Pappa* answers with a smile. After a few days, I stop asking. *Why hasn't Pappa purchased the gun yet? When is Pappa going to avenge me?* I keep on wondering.

In his wisdom, *Pappa* knew he should rather let the *spooktrein* and the bumper cars be, until the day I could face this enemy myself. You see, *Pappa* knew the real enemy was neither the *spooktrein*, nor the bumper cars. The name of the real enemy had only four letters:

F-E-A-R.

*

I feel so powerless here, on the other side of the world. How can I be useful to my parents? Wait! My dad's sisters! Maybe they haven't heard yet. *Tannie Kiewiet*'s phone is just ringing. *Tannie* Mini confirms that Lin has already phoned her and that she is on her way to my dad. I spend the rest of the evening talking to other family members.

My spirit senses that other sis is having difficulty dealing with *Pappa*'s accident. I send a SMS to check on her.

7:02 pm: Other Sis: "I had to calm myself down before I came to the hospital. I felt lightheaded and nauseous when I saw all the blood."

I decide to give her a call, and we chat for a while. Lin left the EU to give other sis and *Ma* a moment with *Pa*, since the hospital didn't want too many people in the EU. While they were busy stitching him up, some of the stitches snapped. Blood spurted from the wound! She is unable to provide me with any more information. Still, she can confirm that *Pappa* is unable to remember anything recent and that he keeps on asking the same questions over and over again.

8:38 pm: Me: (voice note): "*Mamma*, my heart is there with you. I am sure that you are still processing the shock since you had to be strong for *Pappa*'s sake. I know he is going to be okay. I truly regret not being able to be there in person and offer you support. Please remind *Pappa* every so often that I am sending him my love, and do not forget to replay my messages as often as possible. I love you both so much."

8:45 pm, I send another follow up voice note to Lin. She answers straight away. Other sis and my mum are currently with my dad, while Lin is waiting in the waiting room. The brain scan has been done, but they are still awaiting the results. *Pa* cannot remember anything and is very confused. Repeatedly, he asks: "What happened?" Mercifully, he manages to have a sensible conversation with them. His short-term memory, however, *is bokveld toe*!

Lin: "I took a photo of the headwound. It is gruesome! Do you want to see it?"

My tactful baby sister! *Pappa*'s headwound is fifteen centimetres long. After examining him, the doctor was astonished that he survived the accident.

While *Pa* was having his brain scan, my sisters huddled together on a bench in the hallway, praying and speaking life into the situation. Lin reminded the hospital staff that *Pappa* underwent prior back surgery and urged them to be thorough.

9 pm: SMS from Lin: "Brain scan results received. Neither swelling nor bleeding on his brain! *Pa* has been admitted and is being monitored for concussion. What a miracle!"

Me: "Praise the Lord, Sis! I am so thankful! It is wonderful, wonderful, wonderful news! It pays off to stand together in faith." For a moment, I am quiet, savouring the moment. "It is fantastic! God exists outside of time and space. He knew our prayers beforehand and answered them in advance!"

I thank her again for keeping her cool this morning. Lin remained calm when she received that call, despite the shock. She discreetly informed her husband about what had transpired and told the kids that she needed to help *Oupa*. Their dad would drive them to school.

Me: "Sis, please replay my voice note often to *Pappa*! Remind him of my love every few minutes. You said his short-term memory is failing him at present. I fear that he may think that I haven't made contact yet. *Asseblief,* keep on reminding him that I love him ... *Ja*, please send me the ghastly photo!"

Lin relays the great laugh they had, listening to my voicemail. She adds that *Pappa* laughed as well!

I switch off my phone, exhausted but at peace. *Ma* assured me she won't be alone during the night since she has a house full of visiting family members. With thanksgiving in my heart, I enter Dreamland.

*

I wake up just after five thirty the next morning. It is pitch dark outside. Even the birds are still asleep. My very first thought is: my dad!

I grab my phone and press the power button. What's taking the little apple so long to warm up?

After what has felt like hours, I open WhatsApp®. *O, tog*! Shortly after I switched off my phone last night, Lin sent the photo. Clearly, they were not exaggerating! The cut winds from front to back right across my dad's head.

Shaking my head, I close my eyes. *Abba Father, thank you for your mercy and for saving my dad's life!*

4:37 am: Me: "*Sjoe*, it's spine-chilling!"

4:38 am: Lin: "You should have seen it before they stitched him up! *Pappa* is laughing and chatting away now. He truly is okay!"

Me: "Can I phone him?"

Lin: "Yes, he has his phone with him."

I press the call button next to his name on my phone. After what felt like an eternity, I hear his voice on the other side.

"*Pappa*?" my voice sounds anxious, yet glad and filled with emotion.

"*Hallo, my liefling*." *Pappa*'s voice sounds normal as if nothing extraordinary happened. Tears of joy and relief run down my cheeks. We chat for a short while.

Lord, I thought for a moment that I might never hear my dad's voice again. Thank you that – in that moment – You have already saved his life. Thank you, Holy Spirit, for Your expressed Voice from Eternity that assured me that my dad would be okay.

Once again, I hear the Voice from Eternity: "Be still and know that I am God!"[8] It reminds me of yesterday's Voice. *Thank you, Lord, that You are with us every step of the way …*

That evening family and friends visit my dad. They laugh and joke about his short memory (or rather lack thereof).

13 April, Circle of Trust thread:

8:23 am: Other Sis: "Has *Pappa* been discharged?"

8:32 am: Lin: "*Yip*, a few minutes ago, Sis."

9:18 am: *Ma*: "We are on our way to the rugby."

What? Can Pa just remind himself that he has concussion and that he is supposed to take it easy? Surely my nephews will understand if Oupa isn't among the spectators today.

Shaking my head, I cannot help but smile. Typical of Chris Viljoen – he is literally too *hardekop* to stay down!

It is after ten when I walk into my study with my second *cuppa*. On my desk is the forgotten pile of Paces I was supposed to score last night. I open the Afrikaans Pace[9] lying on top of the pile. Suddenly, I realise: God was preparing me throughout the week for yesterday's disaster. The Scripture verse for this week in Niki's Afrikaans Pace, is Joshua 1 verse 9! Thankful, I inform our Circle of Trust:

> "God knew beforehand and equipped us with all the weaponry we needed to overcome the enemy. Hallelujah!"

I found out afterwards that almost all the family who were able to, visited my dad. However, that weekend would stay blurry for *Pa*. He would never recall what exactly happened that morning or who visited that weekend.

Chapter 2
The Caravan

Our final camping adventure just before we immigrated to New Zealand, was at Badplaas. *Ai*, my memories of our favourite camping grounds go way back. I recall photos taken with *Oupa* Piet while we were drinking water from the natural spring; *Ouma* Lenie reminding him not to linger too long in the *rumatiekbad*; me, sneaking away to go look at the origin of the hot spring where boiling water bubbles upwards from deep within the earth; *Ouma*'s warning: "You will burn to death if you fall into that spring!"; fun in the open pool outside.

*

We are gathered outside the old Slipstream[10] around the campfire with red, sunburnt cheeks. It is a warm, sultry summer's evening. *Tannie* Carike's frog choir[11] croaks in harmony with Laurika's cricket choir[12]. I wiggle my nose in reaction to the sharp Tabard[13]-smell. At sunset, *Mamma* always keeps the Tabard stick close by, lest the mosquitos devour us. I become fixated on the red and yellow flames surrounding the crackling hardwood.

A dry twig next to the fire grabs my attention. Smaller twigs branch out from the main twig, making it look like a shrivelled tree. I jump up, grab the twig, and press the one end into the flames until the smaller twigs start to glow. Stretching out my arm into the darkness, I exclaim: "Look, *Pappa*! It looks like a Christmas tree with little lights!"

With small, quick movements of my wrist, I wave the twig back and forth. "Now it looks like fireflies!"

"Kids who play with fire, wet their beds!" cautions *Mamma* with a playful sparkle in her eyes. Her gaze moves to the baby curled up in one of the camping chairs.

"Time to brush your teeth and say goodnight," says *Mamma*.

"*Ag nee*[14]! I don't want to go to bed yet!" I complain.

Pappa figures we still have time for a bedtime story: "Come and sit on my lap. I have a story up my sleeve."

Eagerly I crawl onto his lap.

"Once upon a time, there was a little girl and her brother who lived with their parents on a farm at the foot of the Drakensberg Mountain Range. Her name was Little Rachel de Beer[15]. She was twelve years old. They were rather poor and owned but a few cattle and sheep. Unlike us, they could not afford to go camping here at Badplaas," Dad remarks.

"*Ag, siestog*!" I sympathise.

"On a winter's day in the late afternoon, they were hit by a terrifying blizzard," *Pappa* continues. "Little Rachel and her mum helped her dad to herd their livestock safely into their pens. As her dad started counting the herd, he realised that one calf was missing. Everyone joined in the search, including the five-year-old brother. Clasping his hand tightly, Little Rachel started searching around the house.

"The blizzard intensified. Within minutes, Little Rachel was unable to see their house. She held her brother's hand even tighter.

"'Little Rachel, it is s…s…so c…c…cold! I want to go home!' he cried.

"Little Rachel took off her jacket and helped her brother to put it on. 'There you go, now you will be nice and warm. Come along now, we are almost home,' she comforted him.

"The storm grew fiercer, and the snowflakes blazed onto their bare legs with such force that it caused a burning sensation. 'L…l…little Rachel, my l…l…legs refuse to go further! I want to sit d…d…down!' her little brother sobbed. Little Rachel gave him a pitiful look. Suddenly, they bumped into a large ant hill. She looked around, squinting her eyes in order to search the white canvas around her, but their house was nowhere to be seen.

"'L…l…little Rachel, I am f…f…freezing!' Her brother's voice is but a whimper. She took off her pinafore and folded it around his legs."

A tear runs down my cheek. That poor, freezing little boy! Poor Little Rachel!

"Little Rachel was the daughter of a *boer*," *Pappa* continues, "*en 'n boer maak 'n plan*. She got an idea! She hit the bottom of the ant hill with her little fist, until she managed to create a hole. With her bare hands, she dug and dug until she made a hole inside the ant hill big enough for her brother to crawl into. However, he wouldn't stop crying because of the cold. After covering him with her last few pieces of clothing, she lay down in front of the opening to try and keep out the cold." Another tear runs down my cheek.

"'Hush, little brother, hush' Little Rachel whispered.

"Meanwhile, her parents quit their search for the lost calf. Back at the house, they were alarmed to find that the children were missing. They searched all through the night. They kept on calling their names, but no-one answered.

"Early the next morning, just before sunrise, they discovered the hollow ant hill. Poor Little Rachel was lying in front of the opening, her little body lifeless, frozen to death. Inside the hollow ant hill, they found her brother. His body was cold, but he was still alive. Little Rachel sacrificed her life to save her brother," *Pappa* concludes.

Underneath my cheek, *Pappa*'s shirt is soaked with my tears. My little body shakes as I sob. It was the most beautiful, yet most heartbreaking story I have ever heard. *Pappa* could narrate the most beautiful stories …

*

Let's get back to our last Badplaas camping adventure with *Pa* and *Ma,* just before our emigration: This time we do not overnight in the old Slipstream. Our new dome tent is far easier to erect. We are nonetheless very well equipped with all sorts of camping gear. A heap of wet bedding after a thunderstorm, is not on *my* camping itinerary.

For this reason, the blow-up mattress fits neatly onto a collapsible frame to safeguard our bedding from getting soaked. A snoozy goose down duvet ensures we stay warm at night, despite the thin tent walls. Two sets of colourful, convenient plastic pull out drawers on wheels, bought from Game[16], serve as closets. Dishes, cutlery, and other camping utensils stay orderly and dry the same way.

After packing up camp on Sunday morning and loading the boot and trailer to the brim, hubby decides that we shall not go camping again without a caravan. Apparently, I load up the whole house for one weekend away. *Ha!* He should have been at Mabula's Bush Camp with *Ouma Grappie*[17] during one of our family vacations. She even loaded the huge chest freezer filled with fresh meat from their farm onto the back of the trailer – and I'm only getting started!

Before our caravan dream could materialise, three armed men overpowered us in our home at 3am. Being held at gunpoint for three never-ending hours, convinced us that an LSD-holiday (look-see-and-decide-holiday) in New Zealand may not be such a bad idea.

Without further ado, we hop on a plane. New Zealand steals our hearts! Five weeks are enough to convince us that starting a new life on the other side of the world could be a great adventure – even though we have to cough up five rands for every dollar.

We soon realise that we will have to make do with only half of our household belongings. The houses on Auckland's North Shore that are big enough to fit all our furniture, unfortunately fall outside our budget.

Plans for buying a caravan turn into selling off our camping gear. My son's two-man tent ends up being the only camp reminder to be loaded onto the shipping container. We will have to re-stock on the other side.

Our caravan dream soon evaporates after arriving in New Zealand. The cost of new camping equipment, together with a storage shortage constitutes only two of the many reasons. The biggest damper on our camping spirit is, of course, the absence of a camp-friendly climate (well, according to every newly arrived immigrant, still smelling of Africa dust, of course).

Kiwi's will naturally disagree with this. I guess we can forgive them for that. In a country where four seasons a day are a given, a child of Africa's preoccupation with the sun is rather foreign. Over here, they do everything in the rain, from strolling with newborns to playing rugby.

Twelve years later, the longing for a campground ambience cannot help but to resurface. And so, our caravan hunt begins. (Okay, I'll admit, my borderline obsession with cleanliness when it comes to where I sleep, may have been a major influencer in our choice of appropriate holiday accommodation). We realise very soon that we will get more bang for our buck if we import a brand-new caravan from the UK. To tell you the truth, a two-year-old local caravan would have been a couple of grand dearer. We only need to wait six weeks for our UK import.

As with any goods travelling to our little island from the other side of the world via boat, six weeks soon turned into three months. We manage to take ownership of our caravan by the end of March, with the New Zealand summer almost something of the past.

Like any other brave SA-Kiwi, we shall make the most of our Kiwi-autumn and do this camping thing! Before we know it, a business trip to Auckland turns into a working-holiday long weekend.

Thanks to a bit of luck paid forward by my *Oumagrootjie* Viljoen, the Top 10-camping grounds in Orewa received a cancellation on a beachfront site. (Whenever we ran into a bit of good luck, *Pappa* would always jokingly say: "*Ouma* Viljoen has got my back!" After all, he was her favourite grandchild. Over time, it became a family saying.)

I start packing the day after we took ownership of the caravan. At the break of dawn, we hit the road – *teen 'n slakkepas*! Show me the person who can go faster than 100km/h in Kiwi-land without incurring a speeding fine and a few demerit points? *Nee, swaer*, downunder[18] we are law-abiding citizens who dare not complain!

Late afternoon, we arrive at the Top 10 camping grounds at Orewa beach. We instantly fell in love with our beachfront site. Thanks to modern technology, we only need to unhook the caravan. From here on, we use a remote control to park it right next to the water and electricity supply. Full of confidence, we unfold the awning.

My parents and sister, who are old campers, provided me with a failproof plan to set up the awning. Laying out the annex poles and connecting them according to the instructions first, is key. However, hubby is a descendant of Moses' folk, who would rather wander around in the desert for forty years, before asking for instructions. Before long, he decides that erecting an awning is *easy as*[19] and that I am wasting my time by laying out the poles first.

When I look up, he is already busy threading the awning roof onto the caravan. (Any experienced camper will have a chuckle right about now.) Before long, we discover that the roof poles are too short to fit into the middle pole connectors. Once again, I read the instructions. I compare the numbered poles with the respective quantities as per the instructions. Murphy's Law! Our number 7 poles are nowhere to be found! Before long, we are out of daylight - and out of patience too!

In desperation, I call my son who lives in one of the breathtaking bays on the North Shore. True to his nature, he put his evening plans on hold. He and Harvey (his royal highness the Yorky-Poodle crossbreed) come to our rescue. After examining our number 6 poles, he starts shaking them. Abracadabra! He shakes the number 7 poles out of the number sixes – right before our eyes!

By the time we have sorted the awning, it is already late, cold and the smiles are few and far between. We decide to pretend that it is actually an old-fashioned caravan with neither water nor electricity connections. Thus, we pay a visit to the communal ablutions before hitting the sack. All the cute, modern frills requiring connections will have to wait until tomorrow.

Early the next morning, sunrays slip through the window, glide over my bed, and fill me with renewed courage. Our first day as Kiwi campers! It is time to connect all the gadgets a 2019-caravan has to offer (which would have been a lot easier if hubby wasn't an Israelite).

A few hours later, the wastewater, water supply, electricity and air-conditioning are connected, although the fan regulating the airflow refuses to cooperate. The little red icon, supposedly indicating the hot water supply, remains but a teasing emoji, because the water emerging from the tap, remains ice-cold.

I cautiously open the instruction book once again and read the steps we need to follow out loud to hubby. The Israelite only listens to two sentences before continuing to try and figure it out by himself.

Soothingly, I try again. This time, he listens to three sentences before deferring to his own devices again. Frustrated, he throws his hands into the air. The *verdomde ding* won't work!

Mum would have said: '*Watwou wil*? It has neither mum nor dad now, does it?' Once again, I read the instructions out loud. Finally, I am able to read all four sentences nice and slowly in such a manner that hubby can follow me. He tries again ... Can you believe it? The '*verdomde ding*' works!

Later, I quietly find my way to the beach and send a voice note to Lin. We have a bit of a chuckle over the great test that setting up camp can be for a relationship. A moment of silence follows.

Me: "To tell you the truth, all I want to do is to cuddle up on the bed next to *Pappa.* I want to hold him tightly and soak his shirt with tears of thankfulness that he is still with us."

Lin: "Yesterday, while I was sitting next to him, I broke down into tears a few times, whenever I thought about how different things could have been. At least I am here with him. I can just imagine how you must feel."

Me: "Thank you for being a sounding board. In the contradictory words of Langenhoven[20], I shall slowly make haste and take the long way back to our camp site."

Lin: "'Love you, Sis! I wish you were here!"

Sunday evening, 14th of April I receive a link to *Kruisgenerasie's* morning service on Facebook Messenger from a recent South African immigrant. "I am so sorry to hear about your dad. Lin has just given a testimony during *Kruisgenerasie*'s service."

I open the link and cry tears of thankfulness all through her testimony. "*Dankie, Here!* Thank you, Lord!" I whisper through my tears.

Monday morning 1:05 am: Lin: "The Spirit in you is far stronger than anything in the world."[21]

7:42 pm: *Mamma*: "Every new day is but for the grace of God."

7:53 pm: Other sis: "Amen!"

8:18 pm: Me: "Amen! Therefore, let us rejoice and be glad!"

8:39 pm: Me: "So thankful that our family is still complete, including all the members added by marriage and because of marriage!"

16 April 2019: *M*amma: "When you reach the end of your own strength, you are at the beginning of God's strength!"

17 April: Me: "We are all packed up and on our way home. All the poles have now been colour-coded."

19 April we celebrate Easter. My nephew draws the most beautiful picture of Easter-grace.

21-24 April 2019 we are camping at Papamoa Beach, Tauranga, in the picturesque Bay of Plenty. *Oumagrootjie* once again assured a last-minute beachfront spot, so we can be the first ones to see the ocean every morning!

Between the camp site and reception there is a play park with a jungle gym and trampolines. Niki asks permission to take a stroll to the park.

I am reminded of Buffelspoort – my second-best childhood camping ground, after Badplaas.

*

"*Pappa*, can we go to the swimming pool, please?" I beg. It is clear that the grown-ups are having a *lekker kuier* and that we are going to have some difficulty convincing them to accompany us.

Like most of you, I had two *Pappas*: an everyday-*Pappa* and a holiday-*Pappa*. My everyday-*Pappa* was very strict. Often, when I wanted to ask his permission for something serious, I became so scared working up the courage to ask him, that I burst into tears by the time I opened my mouth to utter the question. My holiday-*Pappa* was always very jovial, lenient, and not nearly as strict as my everyday-*Pappa*. He was also very generous when it came to ice cream and chocolates.

"*Ja*, you can go for a swim," agrees holiday-*Pappa* with a big smile. "Just come tell me every hour that you are still okay," he says in a slightly more serious tone. *And if we drown within the hour?* I wonder – but dare not ask!

*

About thirteen kilometres from Papamoa Beach, is the Mount Maunganui Peninsula – a suburb of Tauranga. It is to the Aucklanders of New Zealand what Durban was to the old *Transvalers*. The most beautiful part of the area, is The Mount itself: an extinct volcano, with the most breathtaking sea views and hiking trails. We climb The Mount every day - choosing a different walking trail each time.

After every trail around The Mount, we spend two hours at the hot pools at the foot of The Mount. Apparently, it is the only hot seawater pools in the world. One would think that with such an extraordinary claim to fame, they would beat Badplaas and Buffelspoort any day.

Alas, a SA-expat needs to make a huge mind shift before he will feel himself at home in Kiwi-land. Not only with regards to the square meterage of his family home, but also regarding both the quality and the quantity of resorts.

Twelve years into it, I am still searching in vain for the typical holiday vibe we used to experience on the beaches and next to the swimming pools at the holiday resorts of my childhood. I am yet to find the swimming pool where a live band performs, while everybody in the swimming pool is singing along 'hoema-le-la-la, hoema-le-la', and clapping their hands together above their heads to the tune of "Four Jacks and a Jill"[22].

No one in any of the Kiwi Holiday Parks show up for *boeresport* to partake in sack racing or tug-of-war. Come evenings, the campfires are few and far between. No one plays concertina and guitar, whilst singing heart-felt folk songs 'till long after the official lights-out curfew. Neither does one hear the well-known *Sewende Laan* tune at half past six in the evenings from various caravan sites.

Here, we learn to appreciate the beauty around us in different ways and to create new campsite memories. The only reason we zip up our tents, is to keep flies, mosquitoes, and rain at bay. Thugs and baboons are very unlikely to invade our tents or camp sites. Spiders are rather miniature compared to their African counterparts and snakes only live in zoos.

I break out in spontaneous laughter at the memory of the pranks the grown-ups used to play on one another at Buffelspoort. During some of my primary school camping holidays, we joined a few other families. Three of the *ooms* recklessly confessed their greatest phobias late in the evening around the campfire. Each phobia would result in a great laugh for the fellow campers.

Oom Diewie was surprised by a frog in his shoe; *Oom* Awie almost got attacked by a rubber baboon spider lurking on his pillow and *Oom* Adri discovered a rubber snake in his sleeping bag. Two of them were rather big and tall. I found it quite odd that two such big, grown men would get such a fright from a *gogga*.

Maybe I have the phobias and the *ooms* mixed up. Nevertheless, the pranks not only resulted in great laughs, but also put some friendships and relationships in serious jeopardy!

*

26 April *Pa* and *Ma* drive down to Mossel Bay for *Pa*'s annual Class of '67 *matriek*-reunion. The '67-club, as they call themselves, are a few old school friends who get together each year, alternating the venues to meet in each one's respective town of current residence. *Ma* mentions that they need to rest after *Pa*'s accident.

7:45 pm: Me: "Sis, how is *Pa* really doing?"

8:27 pm: Lin: "He is okay, Sis. He is still a bit forgetful, but I believe the opportunity to rest will be helpful."

11:28 pm: Me: "Please keep me posted?"

11:33 pm: Lin: "I will, Sis"

11:34 pm: Lin: "Why are you still awake?"

*

27 April *Pa* and *Ma* overnight at Lamon Guest House in Kroonstad. A truck burnt out on the Potch-road[23], and they had to inch forward for kilometres.

28 April: *Mamma*: "Tonight we are going to overnight at Orange River Lodge."

7:09 pm: The Voice from Eternity: "First seek the council of the Lord!"[24] … "Do not fear, for I am with you; do not be afraid, for I am your God. I will strengthen you, I will also help you, I will also uphold you with My righteous right hand."[25] … *"Ik heb u in Mijn handpalmen gegrift."*[26]

28 April *Pa* and *Ma* take on the Swartberg Pass. Needless to say, the adventurers took the exit which meanders all the way down into *Die Hel*. Growing up, I explored almost every mountain pass in South Africa with my parents. Whenever *Pappa* could choose between a thirty-minute route to reach our destination, or a three-hour long mountain pass as alternative, the pass would always win. The Swartberg Pass in the Klein Karoo[27], with the exit going all the way down to the Gamkaskloof – or The Hell – is at the top of the adventure list! Not to mention the beautiful Bainskloof Pass, with the overhanging rock, in the Boland between Wellington and Ceres. Over the years, we have taken several photos posing next to a car parked right underneath that overhanging rock! The lush vegetation of the Magoebaskloof Pass reminds one of the picturesque roads to Bay of Islands. The Long Tom Pass in Eastern Transvaal (now Mpumalanga) includes spectacular waterfalls. My favourite pass, I have to admit, will always be the Prince Alfred Pass through the Knysna Forest. (Fine, I'll admit - *Tannie* Daleen[28] played a huge part in making this pass my favourite!)

29 April *Pappa* and *Mamma* spend the night in Dana Bay, visiting *Pappa*'s youngest sister. *Ma* informs us that *Pa* is very tired. The first few months of 2019 were exhausting.

30 April 2019: First school day of the second term for Niki and me. Much to my amazement, even Maths went smoothly.

2 May: *Mamma* sends a prayer on Circle of Trust.

6 May *Pa* and *Ma* overnight at Bundu Inn, eighty kilometres from Kroonstad.

8 May 2019: My *hartkind* gets chosen to play for the Pumas junior rugby team.

10 May 2019: Niki turns thirteen!

Lin and I talk about the challenges her little girl experiences at school.

25 May 2019: Christiaan turns 10! *Pa* and *Ma* spend the night in Pietermaritzburg.

Towards the end of May, the shareholders of our firm go on a three-day boat cruise.

By now it is way too cold to go camping. But our caravan may have a much shorter hibernation period than it expected.

Chapter 3
The PSA-Count

During our cruise, I decide that it is time to bring my parents over for their promised birthday visit. *Mamma*'s birthday is on the 4th of October and *Pappa*'s on the 20th of November.

Shortly after our return, I phone them to discuss travel options.

3 June, 2019: *Pappa* sends me an e-mail:

Hello, my dearest child.
Pappa trusts that you are doing well.
You must always remember that the special place you hold in my heart is yours exclusively.
Pappa needs to share my heart with you today.
I pray that you will understand what I have to tell you.
Firstly, you need to know that I was – and still am – so excited about the possibility of visiting you. There are a few important factors I need to consider carefully.

Pappa shares all the financial and practical implications of a four-week vacation away from the business.
He concludes his e-mail:

I suggest we pray that the Lord will enable us to visit you over December since this is the only period we can afford to close the business for four weeks. Time will tell whether finances will allow it.
Pappa hopes and prays that you will understand our dilemma.

All my love,
Pappa

I cry my eyes out. It is not as if I wasn't already aware of most of the confidential information Pappa shared with me, but I desperately want my parents here with me for a proper vacation!

True to my nature, I do not send a reply in my emotional state. I lay my burden at the feet of Jesus and have a sleep over it.

At the break of dawn the next morning, I send *Pappa* my reply:

Hello, my dearest *Pappa.*

Ai, Pappa ... believe me, I wanted to buy you and *Mamma* two tickets directly after your accident and invite you to come and stay with me for three months for rest and recuperation. However, I knew it was neither practical nor financially possible for you and *Ma* to stay away from your business for so long. I also realised that it was important for you to be close to your doctors for follow-up appointments. Obviously, I cannot wait to see you again. I am really looking forward to having you and *Ma* here – just for myself – for a little while.

I know all too well how easily a business can suffer if the boss is out of the office – let alone out of the country. Even hubby, who conducts most of his business electronically, must drive to the Auckland office at least twice a week. Whenever he goes overseas for a few weeks, he has to play serious catch-up and sort out all sorts of hiccups when he returns. Your business is even more dependent on you being there in person. Therefore, I understand – even though the tears stream down my cheeks while saying this.

The past year the Lord has been teaching me to trust in Him. We are trying to get back to Auckland, but until now, the Lord wanted us down in Hamilton. Every month the latter is quite clear when we look back and realise how we were able to bless people around us and how they, in turn, were a blessing to us.

Our house has been on the market for about five months now. We made it a matter of prayer from the start. We can do whatever is humanly possible with the marketing, but God is still in charge. Therefore, according to God's pre-ordained will, we believe the house will sell at the right time, to the right person. I also need to trust His timing in various other areas and believe the same goes for your visit.

Some great benefits of a December visit: *Pappa* and *Mamma* can celebrate Christmas with us; the weather is better – especially if we want to go camping; all of us are on holiday then, and hubby has more free time, too. The only problem is that plane tickets are expensive that time of year. This is then the matter we need to bring before the Lord and ask for His supernatural provision.

How are you, *Pappa*? I am very concerned about your health, and I truly hope you are taking it easy. May the Lord bless you financially; may His blessing be upon everything you do; may you find favour with God and men, and may He bless you with health and many more years of living a good life.

All right then, as Tommy Dell's song goes: 'I'll sign off now before I start to cry,' (Duh-hu! The keyboard is already soaked!). 'May God ride with you. Ten-four and ... God bless!'

With all my love,

Die een wat Pappa eerste liefgehad het.

Pappa responds without delay:

Hello, my *Liefling.*

Baie dankie, for your kind understanding and beautiful reply. Be assured that my heart was deeply touched, resulting in much heartbreaking longing. I realise my time is running out, and I believe with all my heart that our Heavenly Father will allow me one more opportunity to see you and one more chance to hold you tightly to my chest.

It is well with my soul. I am blessed beyond measure with a wife and children who love me dearly. It is sad to experience how my strength is not what it used to be. Even long road trips – which always 'wowed' me, are becoming increasingly difficult. However, I do enjoy it still. There is a special togetherness between me and *Mamma* during these trips.

Fortunately, the Lord blesses me with ample health to run our business successfully, which gives meaning to my life. I, for one, will not be able to sit at home, follow *Mamma* around and maybe peel some beans for dinner. That will drive me nuts! (It will drive *Mamma* nuts too!) I know I still have some work to do in God's great

Masterplan. Hopefully, we shall have enough opportunities to appreciate each other's fellowship.
Always remember: "Tropic skies are jealous of your blue eyes – Jealous because you're mine."

With all my love,
Jou verlangende Pa.

Rivers of nostalgia run down my cheeks.

*

I cannot remember the first time *Pappa* told me the story behind my name. But I have known the story for as long as I can remember.

One evening, while *Mamma* was still pregnant with me, *Pappa* was listening to a vinal record of Bing Crosby. The song *"Sweet Leilani"* touched the strings of his heart:

Sweet Leilani, Heavenly flower,
Nature fashioned roses kissed with dew (Leilani, Le)
And then she placed them in a bower (Leilani, Le)
It was the start of you.
Sweet Leilani, Heavenly flower,
I dreamed of paradise for two (Leilani, Le)
You are my paradise completed (Leilani, Le)
You are my dream come true.
Sweet Leilani, Heavenly flower,
Tropic skies are jealous as they shine (my sweet Leilani, Le)
I think they're jealous of your blue eyes (oh, lovely Leilani, Le)
Jealous because you're mine.[29]

While the song was playing, in his mind's eye, he saw a little girl with blond hair and blue eyes. She looked up at him, smiling and saying: "*Pappa*!"

He got up, walked over to *Mamma* and told her: "The baby you are carrying is going to be a little girl. Her hair will be blond, her eyes will be blue, and her name will be 'Lijlanie'."

Mamma's mum, *Ouma Grappie*, was the third child of Dutch immigrants. They emigrated from Holland to South Africa. Prior to that, *Oupagrootjie* was a missionary in Angola. After a bad bout with malaria, he and *Oumagrootjie* settled in South Africa. *Ouma Grappie*'s Christian names were "Lijsbert Schaafsma". Thus, the Dutch spelling of my name.

Pappa played me the song often. Sometimes, he would play the vinal. Other times he would play the guitar and sing it himself, but mostly, he played it on the piano. His touch became gentler than when he played other songs. One could sense how he poured his whole heart into the piano keys.

*

8 June 2019: Me: "*Sjoe*! The flu has me 'man down' this week! *Nou ja*, a *Mamma*'s 'man down' still includes washing, cooking, cleaning, homeschooling and - in the words of the emperor in 'The King and I' – *et cetera, et cetera, et cetera*.

"I know *Pappa*'s symptoms are similar to mine, so *Mamma* and I decided we both have Lin's longing-sickness." I add a wink emoji.

Time and again, when Lin had to attend a school camp or *veldskool* – and especially during orientation in her first year at university – she would become very ill. The illness would clear up the moment she returned home to *Pa* and *Ma*.

14 June: Lin has a dream. We are living on an estate with a panoramic view - and she is here with us!

I battle with flu and let Aaron know that I really miss my dad.

6:02 pm: Lin: "Sis, *Pa*'s PSA-count is alarmingly high! It's 30 instead of the desired 4. Let's pray that this obstacle will become his testimony."

Me: "Amen."

Without avail, I start interceding for *Pappa*.

Circle of Trust-chat:

10:09 pm: Other Sis: "Lij, your other son shaved his beard!"

10:18 pm: Lin: "Isn't he supposed to shave according to the school rules anyway?"

10:20 pm: Other Sis: "Until now, he was still too young."

Lin: "Sorry! I am with you now! *Sjoe*, I actually cannot comprehend it!"

15 June: 6:05 am: Me: "He is growing up way too quickly. Niki went for her first facial yesterday. When she shaved her legs for the first time, she stopped halfway through, compared the shaven leg with the unshaven one and said: 'These long hairs are not my style!'"

Everyone: "LOL!"

16 June: Me – Voice note: "Today, I am truly thankful that we have the privilege to say, "Happy Father's Day!" to *Pappa.* So many of my friends have already lost either one or both of their parents. Let us take a moment today to thank God that we still have both our parents."

During my quiet time on this specific Father's Day, God gave me the following Word: "And whatever you ask for in prayer, having faith and really believing, you will receive"[30]. I grab hold of this promise God is extending to me and cling to it for dear life. I write '*Pappa* PSA' in ink next to this verse and cross out the PSA.

Without hesitating, I share it with Aaron so that we can stand in agreement in prayer.

The next morning, Jesus draws me closer again and whispers in my ear:
"Whatever you ask in My name, that I will do, that the Father may be glorified in the Son. If you ask anything in My name, I will do it!"[31]
Aaron's spirit once again resonates with mine, and the attack increases in intensity.

Two days later, an absent-minded woman fails to give way to oncoming traffic from the right. The blood of Jesus covers His warrior: Niki and I are okay – our car is not.

Three days later, I celebrate my forty-sixth birthday and my nephew his tenth.

24 June: Lin: "This is what happens when Phillip doesn't show up for work: we need to haul the firewood and start the fire from scratch ourselves. Middelburg-temperatures are minus one and eighteen."

Me: "*Agge nee, man*! Toughen up! Last night the temperature over here was one, and the maximum for today is fourteen – and nobody is hauling firewood for us!"

25 June: 12:11 pm: *Mamma* sends a prayer via WhatsApp.

Me: "Why are you not asleep, *Ma*?"

Ma: "You are keeping an eye on me, eh?"

Me: "Yip, someone has to!"

Ma: "Love the technology. You are so far away – yet, with the little mobile - so near."

7:46 pm: Lin: "*Sjoe*, you were chatting away last night!"

Me: "Somebody needs to keep an eye on *Ma* while she is gallivanting in cyberspace!"

26 June: Other Sis: "Peace is what I leave with you; it is my own peace that I give you. I do not give as the world does. Do not be worried and upset; do not be afraid."[32]

During a phone conversation with *Pappa* on 1 July, I once again shared John 14:13-14 with him: "And whatever you ask in My name, that I will do, that the Father may be glorified in the Son. If you ask anything in My name, I will do it!"[33]

Pappa concurs: "*Ja, my kind*, if it is the Lord's will, He will heal me."

It feels like a cold fist clenches my heart as I realise: my dad is ready to go home. Whether he lives or dies – he is happy either way, for he knows where he is going. My spirit revolts. I can recall *Ouma* Lenie's words clearly: "*Soos die Here wil.* As the Lord wills" – not *if* He wills. *Pappa*'s mum never said: "if the Lord wills", but always "as the Lord wills".

For days I wrestle with the Lord about this. Whenever I wrestle with God regarding a certain issue, He usually answers me threefold: through another person, His Word, and a spiritual revelation. One Sunday, while I am listening to a sermon, the preacher declares that we do not have to wonder what God's will is if He has already clearly revealed it in His Word.

During my quiet time, the Holy Spirit picks me up and puts me on His lap. He revisits the touching story of the man who suffered from a severe skin disease. The man fell at Jesus' feet and begged Him: "Lord … if You are willing, You can heal me.[34]"

I experience a revelation from the Holy Spirit and hear the story being told like never before. I can hear Jesus joyously and lovingly call out: "I *am* willing! Of course, I am willing!"

"Jesus reached out and touched him."[35] – and the man was healed immediately. Jesus whispers in my ear: "I *want* to heal your *Pappa*. Therefore, you can approach the Throne of Grace boldly."[36]

On July 2nd, I share this with *Pappa* in a voice note.

Mamma informs me that *Pappa* plans to drive to Pretoria by himself to meet with the urologist so that she can man the workshop. I do not like the idea.

"Can't Michael accompany him?" I ask. *Mamma* thinks it a sensible suggestion and says she'll ask him whether he could. Relief washes over me like a welcome breeze when I read her SMS on Circle of Trust:

> "So happy! Michael will accompany *Pa* to Pretoria on Wednesday."
>
> Me: "Oh, thank goodness! *Mamma* and I really didn't want *Pappa* to drive alone to Pretoria. *Lekker* bonding time for *Oupa* and his first grandchild."
>
> Other Sis: "It's a pleasure, *Mamsie*. Yesterday Michael drove past a horrific accident on his way home from Potch. So thankful for God's faithfulness! Our Protector is not asleep! Thank you for the Blood of Jesus!"
>
> 3 July: Other Sis: "And the devil saw me with my head down and thought he won. Until I said: 'Amen' and stood up."[37]
>
> Me: "When God's warriors go down on their knees, the battle is NOT over; it has just begun!"[38]

The Voice from Eternity calls me up close. I climb onto His lap. He takes my hand and walks with me through the acts of the apostles two thousand years ago: "By faith in His name, His name has made this man strong, whom you see and know. So the faith that comes through Him has given him this perfect health in front of all of you ... let it be known to all of you ... that by the name of Jesus Christ the Nazarene — whom you crucified and whom God raised from the dead — by Him this man is standing here before you healthy."[39] I look up, close my eyes, and raise my hands. *LORD, Lord, I believe it with my whole heart!*

A little while later, I phone *Pappa* to wish him the best with the doctor's appointment and to pray travelling mercies over him. During our phone conversation, I remind him of the words of Joel Osteen: 'Praise precedes the victory!' "*Pappa*, I thank the Lord in advance that you are in the process of healing, as Jericho's walls fell because of the rejoicing and not the other way around!"

4 July: *Ma*: "Misunderstanding at the doctor's rooms. When *Pa* arrived there, they informed him that the appointment is for July 10. They couldn't even confirm the time of the appointment. They are so *deurmekaar*!"

Me: "*Ai toggie, Mamma*! Well, now we have one week extra for spiritual warfare and to praise God daily for His miraculous healing power. Come the tenth, this team of warriors will be ready to crush the devil!"

"I have given you authority to trample on snakes and scorpions and to overcome all the power of the enemy; nothing will harm you."[40]

7 July: *Ma*: "Satan's target is your mind, and his weapons are lies. So, fill your mind with the Word of God."[41]

8 July: *Ma*: "Let your faith be bigger than your fear."[42]

10 July: Another doctor's appointment to attend.

Me: "Good luck for today, *Pappa*."

Pappa: "*Dankie, my lieflingkind*."

Later, we receive more bad news: the cancerous tumour is too big to remove, but *Pappa* needs to go for a biopsy.

11 July: Me: "Faith takes us beyond ourselves to where God is."[43]

12 July: The Lord sends me another message via Joel Osteen:

“Let us drain the miracle working power of Jesus Christ right out of Him!”

Thursday, 26 July, *Pappa* undergoes a biopsy.

Friday morning, just before *Pappa* is released from hospital, I get a chance to speak to him over the phone. *Pappa* sounds all good and positive. A few minutes later, I receive an SMS: “Yippee!!! I am going home!”

I do a happy dance and wipe away a tear...

Chapter 4
Andante, Andante[44]...

It is early, Sunday morning. Everything is still asleep – the birds and the sun, too – for it is still pitch dark outside. After *Pappa*'s excited message about being released, I haven't heard anything from them. I feel around in the dark until I find my phone and hold in the power button. After a few seconds the little white apple icon appears on the black screen. While I'm waiting for the mobile to start up, I switch on the kettle and peel an apple. Like every other morning, I check my WhatsApp® messages first. Thank goodness! There is a voice note from Lindi.

Suddenly, the sweet apple gets stuck in my throat and turns bitter as[45]. Lindi's voice sounds fragile. Just before the doctor released *Pappa*, he informed him that it was evident during the biopsy that the cancer had already spread to the surrounding bone and glands.

My soul feels crushed. Why didn't *Pappa* inform me immediately after he heard the news? I want to phone him, but what am I to say? *Pappa* and *Mamma* are probably still trying to process it themselves. It feels as though my intestines have been ripped out and thrown onto a careless pile on the ground. My husband, son, and brother-in-law are still in Wellington, where they watched the All Black-Springbok rugby match. They will be flying back to Auckland later today. Niki and I are alone at home. Just as well, for I lack the words to share the news. My legs and arms feel paralysed.

Well, paralysed or not, I have to go to church today. Maybe I'll find a piece of driftwood to hang onto within the community of fellow believers. Meanwhile, Niki has woken up. She calls me so we can have coffee together. Fortunately, she is still half asleep and doesn't ask too many questions. I glance at my watch, pretend to be alarmed when I see the time and excuse myself to go and take a shower.

In the shower, safely hidden behind two glass doors, with hot water running over my body, I allow the heartache to burst forth into rivers of tears. Raw sobs tear through my chest. Tears and water mix and pour onto the shower floor. I rest my head and hands against the glass and slowly sink down towards the floor. I crawl onto Jesus' lap and weep.

After a while, my sobs grow quiet, but my tears won't stop flowing. My lips are searching for words to pray, but my brain, which is supposed to send the words to my lips, cannot seem to allocate any sensible vocabulary. The only words my brain seem to remember, are from a song from last night's movie:

> "Andante! Andante! Tread lightly on my ground!
> Andante! Andante! Oh, please don't let me down!"[46]

Over and over, I sing the words, weeping. I get out of the shower mechanically and dry myself - my last bit of energy washed away down the drain. I fall flat on my face before God's feet and desperately grab the seam of His cloak. My clenched fists hold onto it for dear life – as if it is my only source of rescue – because it is!

Crying and begging, I sing softly and with a broken heart:

> "Andante! Andante! Tread lightly on our ground!
> Andante! Andante! Oh, please don't let us down!"

As I repeat these words over and over to our Heavenly Father, the words change to:

> *"Yehova Rophecha, Your healing I proclaim!*
> *Yehova Rophecha, my dad! In Jesus Name!*
> *Yehova Rophecha, let healing blessings flow!*
> *Yehova Rophecha, in You our faith now grows!"*

Whenever I intercede for something specific, I start by calling the Lord upon the specific Names of Provision we find in the Bible. By the time I get to my request, I already believe that He can do it because I have just confirmed that He can by calling upon those

Names. "*Yehova Rapha/ Rophecha*", the Hebrew for "The Lord Who Heals", is the only Name I can call upon now.

The drive to church is a bit of a haze. Mechanically, I sign Niki in at Kids' Church and find my way to the auditorium. I try not to look around too much. Some fellow church members greet me, and I nod briefly with a forced smile. I make my way to the pew as quick as I can. Relieved, I sit down and bow my head. To make small talk will take too much effort today. Without words I just sit before the Lord – empty and dumb.

While I am sitting there waiting, the Voice from Eternity whispers in my ear: "Healing Board."

Lord?

"Make a healing board for your dad."

After the service I rush to fetch Niki from Kids Church. As we get into the car, I pray: "Lord, I have no idea what I am supposed to write on this 'healing board', but I will do what You told me to do." Acting in obedience, I turn into the parking lot outside the $-store on my way home and purchase a large white piece of cardboard.

Until now, I was unable to utter the words: 'My dad has cancer.' When that big letter C touches your heart, it becomes something unpronounceable. It is as if someone steals your breath every time you try to say it.

As I pull into our driveway, Niki asks permission to watch TV. With a sigh of relief, I agree and flee into the safety of my study. My hands are shaking as I lay down the piece of cardboard in front of me. Inhaling deeply, I place both hands on the empty poster. I bow my head and pray: "Lord, like Paul, I stand in front of Your Throne of grace today, and I claim to know nothing, "…except Jesus Christ and Him crucified!"[47]

Instantly it hits me: I must address this enemy! As long as I am too afraid to call it by its name, I am giving it power over me. Thumbling through my stationary drawer, I find a thick, red vivid[48]. I take a deep breath and feel how the turmoil inside me quietens. With a strong, sure grip, I write the words in big, fat, red letters in the middle of the white poster while saying it out loud:

'Cancer, you cannot have my dad!'

Supernatural power takes hold of me – like water slowly building up and then suddenly bursting forth through a crack in the surrounding rocks. *This enemy has now been addressed and needs to be destroyed!* my spirit exclaims.

I search through my scrapbook alphabet letters. It doesn't take too long to find the right size capital letters. With determination and conviction, I glue the embossed words CHRIST WILL HEAL, over the written words 'Cancer you cannot have'. Now the sentence reads:

CHRIST WILL HEAL my dad!

Okay, Lord. The most important part is done. The enemy no longer has a hold on me. I addressed it and declared Your Word over it.

"Give the poster a name," summons the Voice from Eternity.

Once again, I search through the alphabet letters. After a few minutes of measuring, sorting, cutting and pasting, I stand back and have a good look at my handiwork.

MY DAD'S HEALING BOARD

For the first time today, a smile starts to linger at the one corner of my mouth. Like cool, refreshing waves, rolling slowly over the burning sand in order to soothe it, the realisation washes over me: I first had to identify my fear before I could believe in the heading:

MY DAD'S HEALING BOARD.

Joel Osteen's credo echoes in my ear: "Praise precedes the victory!" Underneath the heading and slightly to the left, I write:

"Hallelujah! Praise precedes the victory! Jericho's walls fell because of their shouts of joy and not the other way around."

"I raise a hallelujah![49]" I exclaim in obedience. Fortunately, Aaron and I already have a few songs in our warfare arsenal. One after the other, I play the songs. Thanks to the ever-willing YouTube, I have no difficulty finding the lyrics to each of my warfare songs. While it's printing, I turn up the volume and sing along even louder. The Holy Spirit takes over and fills me with resourceful ideas.

Before long, the lyrics to "I raise a hallelujah," "I will be confident,"[50] and "Whom shall I fear"[51] have found their way to the healing board. I have attached them with colourful strings and golden thumbnails. During my dad's first visit to New Zealand, he bought me a silver chain with a music note pendant just before he flew back home. A few years later, hubby returned from one of his business trips with some Christmas gifts from my parents. This time, I received a silver chain with a cross pendant from *Pappa* and *Mamma*.

I attach the lyrics to "I will be confident" and "Whom shall I fear" to the two silver chains and pin them to the board. The other three strings with lyrics are decorated with two golden crosses and a cross made from New Zealand paua shell, respectively. I realise I need a few more warfare songs to complete the healing board, but I trust the Holy Spirit to advise me accordingly, in due time.

Pensively, I scroll through our WhatsApp®-chat thread and find a few more spiritual weapons. Fully dressed in our spiritual armour, Aaron and I start praying for a miracle like never before:

"Because when God's warriors go down on their knees, the battle is NOT over. It has just begun!"[52]

A picture with the above credo has been on my Facebook profile well before my dad's diagnosis. I print it out and paste it onto the healing board.

Another few weapons from our WhatsApp® arsenal find their way to *Pappa*'s healing board:

"For we wrestle not against flesh and blood, but against principalities, against powers, against the rulers of the darkness of this world, against spiritual wickedness in high places."[53]

"Faith takes us beyond ourselves to where God is."[54]

I step back and study the board with all the Scripture and cries of warfare. In a loud voice, I concur:

"Therefore, I declare in the Name of Jesus Christ: Cancer, you cannot have my dad! CHRIST WILL HEAL my dad!"

Determined, with unwavering faith, I personalise the Scripture Jesus has already given me. I edit Acts 3:16 to read as follows:

"By faith in His name, His name has made this man – Chris Viljoen - whom you see and know, strong. So the faith that comes through Him has given him – Chris Viljoen - this perfect health in front of all of you."

Before long, I would proclaim this verse from the rooftops!

Acts 4:10 and Romans 8:11 also obtain supernatural power through my personification:

"Let it be known to all of you ... that by the name of Jesus Christ the Nazarene – whom you crucified and whom God raised from the dead – by Him this man – Chris Viljoen - is standing here before you, healthy.[55]

Jesus' cry, confirming his willingness to heal, gets added to my healing board with calculated precision and unstoppable faith:

"'Lord, if you are willing, You can make me clean.' Reaching out His Hand He touched him, saying: 'I am willing! Be made clean.' Immediately his disease was healed."[56]

Lest courage evades me when the enemy sneaks around the corner to threaten me, the Voice from Eternity reminds me: "Look, I have given you the authority to trample on snakes and scorpions and over all the power of the enemy; nothing will ever harm you!"[57]

Slowly and thoroughly measured, I read each phrase and Scripture written on the healing board, out loud. "*So ja*! Thank you, Lord."

I grab my mobile and take a picture of the healing board in its entirety. Then I zoom in on the individual credos, Scripture and warfare lyrics and send all the pictures to *Pappa*.

> "*Pappa*, the doctor's diagnosis is the finding of a mere mortal. It is not what is written in the heavens! In the Name of Jesus, I cancel it and speak life over you: healing - not illness. Thank You, Holy Spirit, that You are interceding with mortifying groans on our behalf, since we do not always know how to pray."[58]

"I will be confident" starts playing in the background. I turn up the volume and declare out loud:

> "He made me a promise. He gave me a calling. I know that He's able to keep me from falling. I focus my hope. The past is gone. The moment to prove the power of God has come! In this I will be confident! The time is now; the odds are long; the haters are loud; the pressure is on. But devil, you messed with the wrong one! I got my sling, I got my stone, and in the name of my God I will overcome! In this I will be confident!"[59]

Exhausted, I fall on my knees! As if from afar, Brandon Lake's voice sounds from my playlist:

> "Mountains are still being moved. Strongholds are still being loosed. God, we believe it, yes! We can see that wonders are still what You do. Bodies are still being raised. Giants are still being slain. God, we believe, yes! We can see that wonders are still what you do."[60]

Sela. Amen.

Chapter 5
The Faith of Chris Viljoen's Daughters

I am unable to speak to *Pappa* and *Mamma* again over the weekend. Their house is filled with family and friends from all over – almost as if they went there to say goodbye?

Aaron: "There was no end to my tears this morning!"

Me: "When you went to bed, it was my turn to open the flood gates. The thought of having coffee with my family as with every other morning, was unbearable. While they were drinking coffee, I cried uncontrollably in the shower."

Aaron: "At least it feels as if the shower is washing away some of the tears. I love your healing board. Thank you for sending through that Scripture."

Me: "When I feel paralysed with fear, I take my stand in front of that healing board and declare out loud everything written on it. It strengthens me."

Aaron: "We will carry each other through this. When you are weak, I shall be strong and whenever I am weak, you will be strong."

Me: "I wish we could hold each other in an embrace! If only I could curl up on the bed next to *Pappa* and hold him in my arms!"

Aaron: "I know it's hard on you being so far away. I miss you so!"

Me: "And I you!"

Aaron: "I am unable to utter those words …"

Me: "No worries, I have already done that. Now those words have been nailed to the cross."

Teatime, I find myself a cozy spot in the weak winter sun and nibble on a snack while staring at the clouds. Memories take me back to my schooldays.

*

"*Pappa*?" I call.

There he is, lying flat on his back on the lawn. His knees are bent, his feet on the ground.

"Hey, my Rosebud. Come and lie down beside me," Pappa invites me with a friendly smile.

Eagerly, I obey. Lately, I have seldom had the opportunity to have one-on-one time with *Pappa*. "What are we doing now?" I curiously ask.

"Now we are lying flat on our backs, watching the clouds go by."

"Why?"

Pappa draws in a deep breath and blows it out slowly. "What do you see?" he asks me.

For a moment, there is silence. I turn my head, trying to figure out the picture in the clouds. "Hey, *Pappa*, look!" I call out with excitement. "That cloud looks like Pinocchio!" I laugh.

Pappa agrees. "And there is old Mario, chasing his own tail," *Pappa* jokes.

One after the other, the clouds take on the form of a different creature. We laugh out loud when a creature's nose grows all skew as it transforms into something different …

*

Today, almost forty years later, the clouds look like warriors on chariots with horses, their weapons drawn in warfare.

Positioned where I am, I am facing north. The front line of clouds – or warriors – are moving in a westerly direction towards South Africa. The Voice from Eternity whispers in my ear:

"As you are praying and waging this spiritual war, I am sending angel armies - My angel warriors – to South Africa to fight this thing."

My concentration deepens, and I study the clouds with more intensity. *Sjoe, they are moving rather slowly, eh? At this pace, it will take them forever to reach South Africa.*

The Voice turns His head towards me with an amused smile: "Remember, in the first instance, the clouds up there are moving much quicker than it might seem to you down here. Furthermore, you haven't only been praying for your dad since yesterday, have you?"

My heart fills up with renewed warrior courage. I study the armoured warriors in their different shapes and sizes, which makes me realise that there are different types of demons against whom we need to wage war. Hence the different shapes and sizes of warriors in the clouds.

Behind the warriors, I notice another cloud moving in the opposite direction, towards the east, away from South Africa. Through squinted eyes, I study the form of this cloud. It resembles a pelvic bone. Within the pelvic bone, are smaller clouds resembling glands.

As clear as day, The Voice explains: "While the angel warriors wage this war, the cancer is moving further and further away from your dad."

Jesus! I stand upon this Word! I stand upon it!

3 pm, I put on some make-up – for the first time today. Usually, I would never step out the front door without applying at least a bit of mascara – and I go for very early morning walks!

Later, as the sun sets over New Zealand and rises over South Africa, I phone *Tannie Mini*, *Pappa*'s big sister, who spent the night at my parents' house. Excitedly, I share the vision I had with her. She concurs that she had a dream with the same message the night before. United in spirit, we stand together in prayer for Dad's healing.

Mamma is probably busy preparing breakfast for everyone, so I will rather not phone her now.

Monday afternoon, *Pappa* sends a video message to all his daughters. He is sitting on their double bed. Behind him is the quilted headboard, framing the mirror just above their bed. Blue, crème, yellow and burgundy diamond quilts. *Pappa* upholstered it himself. He is an artist in various fields. For example, he can play almost any musical instrument just by following the tunes and keys in his head. Upholstering is one of his specialities – and he is a perfectionist in this trade.

> "To all my beautiful daughters. I just want to say: Thank you so much! I am indeed blessed to have such wonderful children like you. Thank you for the way you are entrusting me to the Lord in your prayers. Be assured, your prayers are the wings that lift me up, which enable me to approach every day with joy, knowing that I still have so much left to do on this earth.
>
> "When the Lord comes to take me home, I want you to write the following on my tombstone: "The lines have fallen for me in beautiful places; indeed, I have a beautiful inheritance."[61] *Lekker slaap, julle*. Be assured, I love each one of you tremendously. Bye!"

Rivers of sadness flow down my cheeks. Once again, I realise: *Pappa* is ready to go home. He's happy either way.

My spirit revolts, and I fall face-down at the Lord's feet, grabbing His cloak. I remind him how the paralytic's friends dug a hole in a roof in order to get their friend to Jesus. In Luke 5:23, it is written that when Jesus saw the faith of the paralytic's friends, He told him: "Your sins are forgiven ... Rise, take up your bed and walk!"[62]

Like before, I personalise the Scripture and call out: "Jesus, when You saw the faith of Chris Viljoen's daughters, You told him: 'Stand up and walk!' Jesus, Your Word cannot lie! Your Word declares that You want to heal my dad. Even if it is only because You saw the faith of Chris Viljoen's daughters!"

By Tuesday evening, a storm is raging inside of me. Since his diagnosis, I have not been able to have a decent conversation with either *Pappa* or *Mamma*. They had visitors over the weekend, and Monday morning, they were off to work. With trembling fingers, I dial *Mamma*'s number and manage to catch *Pappa* just before he steps out the door. The moment I hear their voices, I start sobbing uncontrollably – especially because I cannot be there with them. *Pappa* speaks a few words of comfort before he says goodbye. The shop won't unlock its doors by itself. For a few moments, *Mamma* and I cry together. We end our conversation with a prayer and speak life over the situation.

Early Wednesday morning, a video message from *Pappa* awaits me. Behind his head, the sky looks grey. In the background, I can see the red pitch of the workshop's roof sticking out and hear the 'zoom-zoom' sound of the sewing machines and the generator.

> *"Hallo, Poplap.*
>
> "I am taking a short break to talk to you. My request is that we take this huge, heavy burden and surrender it at the feet of Jesus once and for all. Then we can forget about it. If not, our faith is not as it should be. His will shall be done, and His will is always, and in everything, the best for us.
>
> "I will keep you posted. When I called the doctor's rooms this morning, they hadn't received the test results yet. I told *Ma* if it was that urgent, the results would have been there already. As soon as I receive any news, I will let you know.
>
> "So, what I am asking of you is to be calm. Plead the blood of Jesus over each other, over me and over all of us. Then we move forward with faith, peace, and a song of praise in our hearts. Because it is well with Pa."

Pappa's words resonate with my soul, but my heart is in pieces! I want to hold my dad in my arms! While my heart and my spirit are wrestling one another, I receive a notification alert:

"A warrior is that woman who gets up despite the enemy trying to destroy her. A woman who declares victory before seeing it. A woman who believes she will receive her miracles because she knows the Lord she serves is alive and powerful."[63]

Aaron and I agree: This will be an amazing testimony!

Chapter 6
Aaron and Hur

The seventeenth chapter of Exodus tells the story of the Israelites and the Amalekites.

> "At Refidim, Amalek came and fought against Israel. Moses said to Joshua: Select some men for us and go fight against Amalek. Tomorrow I will stand on the hilltop with God's staff in my hand.'
>
> "Joshua did as Moses had told him, and fought against Amalek, while Moses, Aaron, and Hur went up to the top of the hill. While Moses held up his hand, Israel prevailed, but whenever he put his hand down, Amalek prevailed. When Moses' hands grew heavy, they took a stone and put it under him, and he sat down on it. Then Aaron and Hur supported his hands, one on one side and one on the other so that his hands remained steady until the sun went down. So, Joshua defeated Amalek and his army with the sword."[64]

Spiritual warfare requires a careful selection of fellow spiritual warriors. Partners in spiritual warfare need to be on the same wavelength and speak the same spiritual language. It requires a person you have a close relationship with – not merely an acquaintance, and more than just a nice lady from church. These partners may not include the friends you socialise with on a regular basis.

From the start, Aaron and I were certain that my dad's battle against cancer was much more a spiritual war than a physical one. We also knew we would have to carry each other through this war.

Aaron's spirit resonates with mine. Paul writes in his first letter to the Corinthians: "The spirits of prophets are subject to prophets."[65] When a revelation I received from the Holy Spirit, resonated with a revelation one of my spiritual partners perceived in her spirit, I could always rest assured that I was not just trying to convince myself of that which I wished to hear.

Aaron and I are in agreement: *Pappa* will be healed! Some days are better than others. Certain days I am carrying Aaron; other days, Aaron carries me. Some days we are physically ill, which requires healing prayers and finding quick remedies for our ailments, for a sick warrior puts up a mediocre fight.

An interesting phenomenon becomes part of my warfare. Lately, I have been experiencing symptoms that come and go. The other day, it was a bladder infection. Then *Mamma* told me that *Pappa* was experiencing some discomfort with his bladder. Last week, my tummy gave me trouble. *Mamma* informed me that *Pappa* really battled with his stomach.

Sometimes, my chest feels so tight that I cannot breathe. The next morning, *Pappa* will report on having a restless night, due to breathing problems. I learn to pray for whatever symptoms I experience until they disappear. Usually, *Pappa*'s symptoms improve at the same time as mine.

Some days, when the enemy is trying to strangle me, I look at myself in the mirror while I'm waging this spiritual war to convince myself of our victory. I discover "101 healing scriptures" on YouTube, and on these extremely difficult days, I put it on continuous play while I'm doing a workout or making dinner - living and fighting for my dad.

*

Aaron: "Today, I feel a bit weak, tired and attacked. My children and I have bronchitis, and my fighter's ohm is on reserve!"

Me: "*Ai*, my darling Aaron, the enemy attacks us, because we continue to throw spanners into its works. Yesterday, I reached a bit of a low. The enemy is trying to paralyse us, but he can't."

Aaron: "I can feel how your wings are carrying me."

Me: "*Pappa*'s healing is already happening!"

Aaron: "Jesus will heal *Pappa*!"

The Voice from Eternity: "You have to pray, believing! Do not doubt!"

Me: "Have a good night's rest, Aaron. I am fighting fit again. Make yourself a hot toddy[66]. (Remember the brandy!) Tomorrow, you'll be all good again."

Aaron: "We are taking turns with the nebuliser before we go to bed. It feels like it is taking forever!"

The Voice from Eternity: "I know you are spiritually exhausted and drained. Hang in there! I will carry you through!"

Me: "Go to bed, Aaron. The Holy Spirit is interceding on our behalf with groans that words cannot express – especially when we ourselves are too tired and dumbfounded to pray. Take care of yourself! You are no good to anyone if you are flat on the floor." I end the message with an emoji of a reprimanding schoolteacher.

Aaron: "I really appreciate you!"

Later on, Aaron's dark night is over, and she encourages me with a message from the Voice from Eternity:

"A warrior is that woman who gets up despite the enemy trying to destroy her. A woman who declares victory before seeing it. A woman who believes she will receive her miracles, because she knows the Lord she serves is alive and powerful."[67]

Me: "Amen!"

*

Just like with Moses during the war with the Amalekites, Aaron is not my only pillar of strength. Hur is here as well. Months before

Pappa's diagnosis, Hur and I were waging spiritual warfare on behalf of her mum.

Hur's mum was diagnosed with brain cancer. Despite the doctor explaining prior to her surgery that she would most likely not recognise her family after the surgery, and that she would have to learn how to walk and talk again, she did not doubt for a second that God would be with her. He sent an angel during the surgery, giving Hur's mum holy water to drink. She told us afterwards that it was the sweetest, tastiest water she ever drank!

When she woke up after the surgery, she immediately recognised everyone gathered around her hospital bed. She was able to walk and talk – even though she initially had to take it slow. A few days later she was her old, friendly self again.

Hur is to me what Aaron, there in far-off Africa, cannot be. She is … here!

Unfortunately, Hur's mum's struggle was not finished yet. The same personified Scripture in Acts, which I sent to my dad, was first personified for and sent to Hur's mum (my New Zealand mum). By the time *Pappa* received his diagnosis, Hur and I were already fighting hard to overcome the war against cancer.

There were days when Hur was crushed due to bad news from the doctors. On those days I was holding her arms up in the air. Other days I was crushed when *Pappa* was the recipient of dreaded results. That was when Hur held my arms high.

Whenever the distance between my dad and I becomes unbearable, I call Hur's mum – just to hear how she is doing. It is ALWAYS going well with her, regardless of hopeless test results, lack of energy or collapsing veins during yet another round of immunotherapy. Her faith is unwavering through it all, for it is built on a foundation of rock. Whenever she speaks, she speaks life into her situation! The moment I hear her voice, I know: All is well with my dad. He will be okay.

*

12 August: Hur: "Today, I am praying for you, your family, your sisters, your mum and dad, as well as everybody who loves your dad. I pray for peaceful hearts, and that this week will be the beginning of renewed strength. God knows us intimately. Therein lies our comfort. He cares for us. In this, we find security. We know our God is greater than any illness or medication. Sometimes, God does bring healing through medical science. Through it all, God remains our hope. Isaiah 40:31 says: 'Those who wait on the Lord shall renew their strength.'"[68]

Me: "Thank you for this beautiful message so early on a Monday morning! I pray Romans 15:13 over you: 'May the God of hope fill you with all joy and peace in believing, that you may abound in hope, in the power of the Holy Spirit.'[69] Together, our families will testify of the wonderworking, healing power of Jesus Christ and how our hope was not in vain."

Hur: "The trust you place in the Lord is indeed the weapon with which you fend off all evil. Your faith is strong; so is your love for God. I thoroughly enjoyed our chat on Friday. After you left, I realised that your eyes still shine with love and your generous smile comes straight from your heart. You have a supernatural positivity, despite the circumstances and even though there is heartache all around us and within us. This will be your testimony, for sure. On the contrary, it is already your testimony. I am reminded of how you shared that your dad is at peace and that he does not want you to be worried. That video he sent you proves that he is okay and is a sure sign that your dad is relying on God completely. Knowing this is comforting."

Me: "*Ja*, it is by grace alone."

13 August: Me: "Hur, thank you for your support and that you are fighting the good fight with me. You are fighting this on behalf of your own mum first, but as sisters in Christ, we are also fighting this together on behalf of our respective parents."

Hur: "*Mamma* was sad yesterday. She has difficulty gaining weight. Treatment can only continue if she gains some weight. She has already lost 14 kilograms! I reminded her that we have put on the full armour of God and that we shall win this war! Thinking of our conversation on Friday, maybe you can offer some dietary advice to assist her in gaining weight?"

14 August: "I received a link to "Chris beats Cancer" on my Facebook News this morning. The first coincidence was his name: Chris – namesake to both your dad and mine! Best of all – he is a Christian. Therefore, a lot of his beliefs resonate with my spirit. He focuses a lot on diet. I will send you the link."

Hur: "God never said our journey will be easy. When I was young, I always thought Proverbs 3:6 meant that children of God will have smooth sailings if they consult Him in everything they do: 'Acknowledge Him in all your ways, and He will make your paths straight.'[70]

"Today I know that even children of God need to conquer mountains. The difference is that if we didn't know God, we would have had to climb every hill without water or a Helping Hand. For us who know the Lord, our paths are straight. Even if it does not appear to be so, His power enables us to climb that mountain as if it were a straight path."

Me: "So true. God never said it would be easy. Thus, He gave us the power to step on snakes and scorpions." *The tune announcing the end of the washing machine's cycle sounds in the background*. "I registered for the "Chris Beats Cancer"-program. As you know all too well, I am somebody who needs to put my faith into action. God created us in His image, barely less than the angels. Therefore, we need to think for ourselves and live out our faith constructively."

Deep in thought, I start taking out the washing piece by piece. I shake it out and drape it as neatly as possible over the clothes airer. Ironing is not really a thing for us immigrants ... *Ai*, Florence[71], if only you knew how much I miss you!

*

Me: "Your mum notified me that she was heading to her next treatment. I assured her that I am interceding on her behalf and sent Acts 3:16 and 4:10 to encourage her. This is the same Scripture I am praying over my dad, for I genuinely believe your mum and my dad are in the process of complete healing. How precious is this Scripture: 'By faith in His name, His name has made this woman, Hur's mum, strong, whom you see and know. So the faith that comes through Him has given her this perfect health in front of all of you ... let it be known to all of you ... that by the name of Jesus Christ the Nazarene – whom you crucified and whom God raised from the dead – by Him this woman is standing here before you healthy.'[72]

"That same power that raised Jesus Christ from death to life, is living within us. That power is busy healing your mum. You are in my thoughts and my prayers are carrying you."

Hur: "Your voice is filled with calmness, with an undertone of burning fire. It may sound contradictory, but Girl! You have a gift. I think you need to start doing audios to inspire and encourage others. What about inspirational devotionals on audio?

"As I listen to your voice while I'm driving, it feels as if a ton of weight is being lifted from my shoulders. Lately, I am also personalising Scripture and songs. It gives life to it. I am learning a lot from you. We are constantly interceding on your behalf as well.

"While I was making my bed this morning, I realised that you have been there for us ever since you found out about my mum's diagnosis. Please phone me day or night if you need me. Jesus Christ is our living hope!"

Me: "I am humbled by your words ... You, journeying with me, means the world to me. Thank you for allowing me to journey with you. I think you know what I'm getting at. It is as if my feet are treading a little lighter because you are lifting my wings. I am going to book my parents' plane tickets today!"

Hur: "This morning I interceded on your behalf, especially. May you experience how my prayers are carrying you. I share in your excitement about your parents' visit. You are going to have the best of times! What a privilege to spend Christmas with your parents!"

*

Me: "Good news and bad news: We received an offer to purchase from a couple who wants to buy our house! On the one hand, I am already popping the champagne cork, but on the other hand my heart is breaking because my time in Hamilton is almost over. The purchasers would like to move in four weeks from now. We are leaving for Atlanta in a week!
"Things are happening really fast now. Still, I am super excited! God told me that things are going to start happening in August and that I will no longer live in this house come 1st of October. We are interested in a house in Auckland. Should this house end up not being the one, I know God has an even better house in store for us!"

Hur sounds overjoyed about our news: "I am so happy for you and pray that you and all your animals will be very happy living on your new lifestyle block. May it be a haven of peace to you. Just think about it! By the time your parents arrive, you would have already moved into your new house - *Deo Volente*[73]!"

Me: "Take note of the people whose eyes light up when you talk about your dream. Those are the ones you keep!"[74]

Hur: "So glad to hear I'm a keeper!" *I can hear the laughter in her voice.*

Me: "Not sure if you went to church on Sunday? I didn't see you at the first service? Dylan Long visited. He encouraged us to get out of the boat in the middle of the storm. Right there and then, the Holy Spirit spoke to me: 'It is rather comfortable to testify after the fact, how Jesus calmed the storm, whilst sitting safe and dry in the boat. However, stepping out of the boat amidst the storm - that shows faith!'
"'So, what if I fall flat on my face whilst walking on water?' Dylan asked at one point. 'Then I know Jesus will catch me!' Dylan proclaimed that he would rather be soaking wet and in the arms of Jesus than high and dry and still in the boat. God wants me to testify now about what He is going to do – before the human eye can perceive it. The Holy spirit is urging me!"

Hur: "We went to the evening service. To think that Dylan had a speech difficulty! Do not let anyone hinder you from testifying, my friend! It is wonderful to experience your journey with Jesus. Your faith is like a bubbling spring that flows all over me, your family and everyone who crosses paths with you."

Tears fill my eyes as her words touch my soul.

Me: "*Ai*, Hur, you only see the best in me. Thank you for sharing so sincerely in my joy, as well as in my sorrow."

Hur: "I am most definitely sincerely sharing in your joy and walking this journey with you. Should you not hear from me for a few days, please know that it is just everyday chores and responsibilities (hubby, kids and my job) that's consuming my time. Please forgive all the voice notes and phone calls while I'm driving, but my car is my happy place where I can talk to you in peace."

Me: "Oh, yes! When I'm alone in my car, it is my happy place, too. Alone time with Holy Spirit."

Hur: "*Mamma* delivered her testimony at church on Sunday morning. She completely forgot to tell you and me about it! Fortunately, we can watch the Facebook video."

Me: "*Ag nee*! I would have driven up to Auckland for that! You and your mum are like shining beacons on my journey. You started this journey ahead of us and lit the guiding torches to show the way. I cannot wait to share a cuppa with your mum, hold her tight, and experience how Jesus restored life within her in person!"

Hur: "My mum is so excited that you are moving back to Auckland."

Me: "How am I ever going to leave you behind in Hamilton? The Lord blessed me with so many true brothers and sisters in Christ here. Well, I won't say goodbye. I will only say: 'See you later.'"

Hur: "I agree! We won't say goodbye. I often visit my mum and sister up in Auckland. I will definitely pop in at your new house. Maybe we'll see each other even more often in Auckland than here in Hamilton! Once you find a *hartsvriendin*, you don't let go."

*

Me: "I put my words into action and e-mailed Jan regarding testifying about *Pappa*'s healing which we believe is already happening. Father's Day is the only Sunday I will be in Hamilton before we move. Jan mentioned that the program is already a bit full. We agreed that I will visit one Sunday after our move. I can either testify during the sermon, or their film team can record it. The recording can then be watched at all the campuses during all the services.

"I wasn't too keen on the recording since I believe a personal testimony has much more impact. However, a recording can reach a lot more people."

Hur: "You can still be at one of the campuses when they play the recording. That way you will be available to chat to the congregation afterwards. I doubt your dad will have any objections."

*

Meanwhile, we are struggling with our ESTA application. It is like a VISA for people visiting the US. Oh, dear! It is a nightmare! Everything is done online, but if the system doesn't like your passport photo, due to your hairdo or unacceptable head to shoulder ratio, it rejects the photo. To top it all, our address hasn't been registered on the system yet and gets rejected time and again. Finally, we figured out that if you persist by typing in the exact same address three times in a row, the system will accept it as your real address. (Of course, we first tried to type it in all the possible different ways we could think of, before trying the exact same way three times in a row!) Nevertheless, we praise the Lord for small mercies!

Another miracle: While I am busy packing for the US trip on the one hand and the *Groot Trek*[75] back to 'Sodom and Gomorra' on the other hand, my little girl finished mathematics all by herself without asking for help once. Praise the Lord for huge mercies!

My darling girl is growing up far too quickly. She is experiencing all the trauma and challenges with her mum. Fortunately, she also shares in the victories and miracles. Hallelujah! Niki sent me a SMS whilst visiting a friend the other day:

"*Mamma*, what other songs did you play during your spiritual warfare? I can only remember two of them now."

My heart was overflowing with gratitude as I sent her the names of the songs. Wherever she goes, she testifies about our journey.

Oh, the sun just emerged from behind the clouds! I think of Hur and quickly share the blessing with her:

"I pray that the sun will break through your proverbial clouds today."

Hur: "Thank you for my sunshine! I'll take that! Now I feel great and ready to face the day."

Me: "I listened to Chris' module on diet today. The anti-cancer diet sounds acceptable and edible to me, but I'm not so sure that it is my dad's cup of tea. While listening to some of Chris' suggestions, I can already hear my dad commenting: '*Ag*, I'd rather die happy and eat whatever I want!'

"*Mamma,* Lin and I will have to be creative when planning Pappa's meals. Small adjustments can make a huge difference.

Hur has a good chuckle over my dad's expected comments.

"Maybe he can drink some carrot juice in between his meals? If he can manage it for two weeks, it will become a habit."

Hur and I continue to share ideas which might be helpful to our respective parents.

Me: "Please pray for me for discernment and wisdom. I want to spend the next few months getting our house into an anti-cancer environment. Not only do I want to try out recipes my dad might like, but I also want to get rid of chemical detergents in and around my home that might be carcinogenic.
By the time *Pappa* returns to South Africa, I want to hear him say: 'My goodness, I have no idea why, but I feel like a whole new person!'"

Hur: "I believe your dad will be so full of energy after his visit that he will never be the same again!"

*

Me: "I have just listened to your mum's testimony! I am planning to host a Christmas Tea in Auckland. It will be wonderful if your mum can share her testimony there. The talks with you and her bless me abundantly! The enemy hates that we are edifying and blessing one another."

Hur: "Indeed! Your encouragement inspires me to slay ten thousand giants! Hur – the giant slayer! During my devotionals this morning, I was reminded of you:

"24 August: 'Dread drains. Faith energises.'
"The Scripture reading is Psalm 55:22: 'Cast your burden upon the Lord (release its weight) and He will sustain you. He will never allow the righteous to be moved (to slip, fall or fail).'[76]
"You never need to drag yourself out of bed. Your faith gives you the energy supply you need. You do not allow the enemy to defeat you."

Me: "Thank you, my friend! Your mum approaches her journey so positively, that it almost sounds as if she is thankful that cancer came her way."

Hur: "She does create that impression sometimes! You know, *Mamma* asked the Lord to use her to bless others. When she received the diagnosis, she immediately devoted herself to honour Him on this journey."

Me: "I think the enemy is livid that none of his attempts defeats us. He probably suffered more strokes than anyone else on earth because of the stubborn persistence and hope of believers!"

Hur: "I can see that! Our discussions these past few weeks have bestowed upon me a double portion of blessings. You make certain to acknowledge Him in every discussion."

Me: "*Ai*, Hur, I miss you already! I will try my best to fit in a coffee date with you before we leave!"

*

Aaron and Hur have me, like I have them. However, I am not their only fellow warrior. There is another Aaron and Hur, forming an Aaron-and-Hur-pillar of support on their other sides. Hur's mum is standing behind Hur and me. With her one hand, she supports our intertwined arms; with her other hand, she supports the intertwined arms – the Aaron-Hur-pillar – of Hur and her sister. Just like my *mamma* is standing behind Aaron and me, keeping our intertwined arms high. Together, these two families battle an intense war against a spiritual adversary who took on the form of cancer.

God is looking down from above onto a network of Aarons and Hurs. They form a beautiful mosaic, part of a bigger jigsaw puzzle. Within each one's agony and struggle, the pearly gates of heaven shine forth, and the bowls filled with the tears of the saints are glistening in His glory.

As we look up, the heavens reflect breathtaking shades of pink, purple, blue, orange, silver and gold when the Holy Spirit and Jesus take their seats next to God the Father, while Aaron and Hur are praying: "In the same way, the Spirit helps us in our weakness. We do not know what we ought to pray for, but the Spirit himself intercedes for us through wordless groans ... Christ Jesus ... who was raised to life - is at the right hand of God and is also interceding for us."[77]

*

Early morning, 4 August, Aaron's friend has a dream. She dreams that she and Aaron are visiting *Pappa* in hospital. Her friend's son accompanied them. Before they entered the hospital, he complained about a little man bothering him. She explained to her son to plead the blood of Jesus over himself. That would provide him with the necessary authority over the pestering little man.

It did indeed cause the little man to disappear. Her son said he would like to pray for *Pappa* as well.

Pappa's hospital room had three bunk beds. He was lying on top in the middle row. "What is bothering your dad?" the friend asked. Aaron mentioned the name of a terrifying demon.

Pappa told them that there was a zebra stallion and a short, brown, fat snake in the hospital. The snake bit the people continuously, yet nobody chased it away. Although *Pappa* did not mention the name of the terrifying demon, Aaron's friend perceived the snake as a fiend that had to be sent back to hell.

A man was sitting in a chair in *Pappa*'s hospital room, close to the window, surrounded by sunbeams entering through the window. The snake bit him over and over on his calf. The nurse sitting next to him, took the snake off him and guided it out the door.

Aaron, her friend, and the boy prayed together for *Pappa*. But before they prayed, Aaron's friend urged her to first plead the blood of Jesus over her and her family and to ask that God's angels will protect them. Together, they commanded the angels in Jesus' Name to fight for *Pappa*.

This dream resonates with my spirit. From the start, the Holy Spirit informed me that this was a spiritual war and that the cancer was demons attacking *Pappa*. We are fighting a spiritual war with and for him. Although he is making use of alternative medicine to fight the cancer and to strengthen his body, it is Jesus Christ who is busy healing him.

Aaron realises it is time to tell her son about his *oupa*'s diagnosis. The ten-year-old warrior is stronger than she thought. Instead of hysterical tears, he retaliates with a powerful prayer of intercession:

"Lord, just like You opened the sea for Moses, please open this thing growing inside *Oupa* and cast it out! The viruses are chasing us, but just like You overpowered the Egyptians by closing the sea over them, overpower this illness and send it back to hell. Oh, Lord, close all the gates so evil cannot come near us."

Sunday morning, our pastor's sermon was about the time when Jesus' disciples were gathering behind locked doors – terrified of the people who crucified Jesus. Pastor drew our attention to the fact that the locked doors could not keep Jesus out. Our fear, doubt, heartbreak, and mourning do not prevent Jesus from coming to us. Jesus meets us right here where we are.

This morning Jesus shared with me His empathy for our hurts. My heartbreak over my dad does not prevent Him from coming to me and meeting me right here where I am sitting – broken, scared and crying.

Me: "Aaron, we most definitely need to pray the blood of Jesus over ourselves. Look at how the enemy is attacking us: first, he attacked our spirits. Our heartache was huge! Then he realised it couldn't break us, because of our strong anchor in the Lord and close spiritual connection with each other and the Holy Spirit. To his disappointment, he realised that a cord with three strands could not be easily broken. Twenty-four hours a day our prayers attack his ears. When your voice grows dim in the West, my voice becomes clear from the East. When I go to bed, you continue in prayer. For this reason, he cannot attack us in our dreams, for we are continuously praying for each other."

I share my dream of the night before with Aaron:

"Last night, hubby was away on business. I forgot to open my Bible at Psalm 23 before I went to bed, as *Pappa* and *Mamma* taught us. I dreamt we were visiting South Africa. We stayed in a holiday unit. Teenaged girls wanted to fetch something from our place. They forgot to lock the security gate after they entered. I looked back and reminded them: 'You didn't lock the gate behind you.'

"We struggled to lock the gate. Out of nowhere, a man whom we knew, appeared outside the gate. 'I am sorry,' I told him, 'I do not care who is with you, but you cannot enter.' Underneath his armpit, a long-barrel rifle appeared. All of a sudden there were four men behind him.

"With sheer willpower I woke myself from the dream. Determined, I got up and started to pray the blood of Jesus over you all and over us. Carmen's song, "The Champion" was playing: 'He has won! He has won! He's alive forevermore! He is risen, He is Lord!'[78]

"It is good that you told your son. Look at what a mighty warrior he is! We need to carefully discern who we allow in our inner circle. As the saying goes: a chain is but as strong as its weakest link. The size of your son's body and the number of years he has been on this earth should not let us doubt his capability to wage this war alongside us. He is *Pappa*'s namesake and believes with the innocence of a child. Jesus said in His word that we should have faith like a child."

Aaron: "My anxiety is gone now. I am experiencing intense peace. No one attacked me in my dreams. Although I did dream that I was bleeding uncontrollably ... Thank you that we can speak from one spiritual mouth and support each other."

*

Everybody in my house is sleeping peacefully at 5am, Wednesday morning, 5 August. I am curled up on the recliner with a cuppa in my hand, fellowshipping with the Holy Spirit around God's Word. He shares the following with me:

“Jesus replied to them, ‘Have faith in God. I assure you: If anyone says to this mountain, ‘Be lifted up and thrown into the sea,’ and does not doubt in his heart, but believes that what he says will happen, it will be done for him. Therefore, I tell you, all the things you pray and ask for — believe that you have received them, and you will have them. And whenever you stand praying, if you have anything against anyone, forgive him, so that your Father in heaven will also forgive you your wrongdoing. But if you don’t forgive, neither will your Father in heaven forgive your wrongdoing.’”[79]

Me: “Often the people closest to us, are the hardest to forgive. We need to search our hearts and be assured that we do not hold anything against anybody. Who are we not to forgive? We must heed against unforgiveness, hiding within the depths of our hearts.”

Aaron: “I realise I need to acquit them of all wrongdoing.”

Me: “*Ja*, I realise I also have to double-check whether I did indeed acquit them completely.”

Sometimes, we need to forgive someone daily. Forgiveness is a purposed choice. It can be so hard that we need to call in the help of Jesus. We may need to take capture the thought of a wrong done to us without delay and make that thought subject to Christ. *Help me, Lord. I choose to forgive them but help me in my weakness.*

I have been experiencing a paralysing pain in my back lately… *Pappa* is suffering from lower back pain…

*

Little foxes are often marauding our vineyards.[80]

6 August: Aaron: “This morning the Holy Spirit spoke to me about the time when Paul preached in the upstairs room in Troas (Acts 20). A young man, sitting in the window, fell asleep while Paul was preaching and fell to his death.”

You see, people fell asleep during sermons even in Paul's days. The consequences were just a little more severe!

> Aaron: "Paul interrupted his sermon and rushed to the boy. After touching him, he calmed the people down and told them the boy was alive. Then he went back upstairs and continued his preaching.
> "In a similar way, the enemy tries to interrupt our spiritual war. He often uses little foxes to try and distract us. Sometimes the little foxes are important matters that require our immediate attention. God reminds us to redirect our focus back to Him whenever we get distracted."

At present, the little fox in Aaron's vineyard is illness in her home. My little fox (apart from my back pain) is our house that needs to get sold. I ask Aaron to pray with me:

> Me: "A third couple, who has shown 'serious interest' in our house, are coming for another viewing. It's been seven months now since we put our house on the market. The weekly commute to Auckland is getting to hubby.
> "I have been begging the Lord the whole day. I cannot bribe Him! Besides, we are praying for God's perfect will in our lives and for His perfect timing. Everything inside me, though, feels it should be happening now!
> "These are the annoying little foxes that I need to focus on right now.
> "Niki and I are also sick again. Before I forget, I sent some Barbies with *Swaer*[81] as well as two books ("Appel Grove Girls" about the girl who was bullied, and "The Girl with no Nose"). *Swaer* will arrange a visit with *Mamma* and take along all the gifts I bought for you guys.
> "Remember, it doesn't matter how many little foxes or fears and illness we have to endure. God walks right through our locked doors and says: 'Peace be with you!'"

The night of 8 August, Aaron's son sleeps through the night without coughing!

One of the most important lessons Chris Wark taught me, was right at the beginning of one of his sessions on how to support a loved one with cancer. He reminded us that we must remember that the journey belongs first and foremost to the person diagnosed with cancer. Sometimes, that person is ready to go home. If that is their wish, we need to respect it.

I had to chew on this one for quite a while. Thanks to my Aaron and Hur, we could chew on this together, encourage and bridle one another where necessary.

One day, Hur and I phoned each other after receiving the same revelation from God. I cannot recall who called the other one first, but God told us: "What if this journey was never about Hur's mum, nor Lijlanie's dad? Maybe this is about something much bigger. These testimonies are going to turn people's lives around." Hur's mum said something similar during her testimony in church. She asked God to use her on this journey to help others.

Another heroic deed of my Hur: Father's Day, when my ill daddy was sitting on the other side of the world, my darling Hur did not go to her dad before saying a special prayer for me. A heartfelt prayer, filled with tears and encouragement – the prayer I could not verbalise that day. A few days later, I received a message: This time, it was Hur's mum who saw her doctor again. The enemy brought them to their knees with the devastating news ... Now it is Hur's turn to lie curled up on God's lap, without any words. She didn't need any, for she knew I was praying for her.

My thoughts go to another Hur. In 2 Timothy 4, Paul informs Timothy that everybody deserted him except Luke. Colossians 4 refers to Luke as 'the dearly loved physician'[82]. Luke accompanied Paul on his mission trips, shared his hardships and persisted with him until the end.

Hur and I are also together on this mission trip. We call it such because we have received a commission to be torchbearers. Torchbearers of whom her *mamma* and my *pappa* are the forerunners.

Dear reader, make sure you have an Aaron and a Hur in your life!

Chapter 7
Hope

Every bone in my body is aching! My back and joints – especially my tennis-elbow! Whenever I am chatting to *Mamma*, I realize that my symptoms are simply imitating *Pappa*'s.

Our fourteenth wedding anniversary is on the 9th of August, and we decide to make a whole weekend of it. The celebration kicks off on Friday afternoon, with an exquisite luncheon and a Couple's Foot Treatment at a Day Spa on the banks of the beautiful Waikato River. Early Saturday morning, we leave for Auckland. After checking in at the *Spencer on Byron*, we are off to enjoy "*Afrikaans is lekker*" in Takapuna with Jo Black as one of the performing artists. This is the first time I hear the song "*Bring die hoop weer terug*". By the time he hit the chorus for the second time, I join in with a loud voice as a tribute to my dad. Jo shares a few words about his dad. His songs – especially "*Bring die hoop weer terug*", "*Skepe*", "*Jy't my gewys*" and "*Vrede Vind*" send a wave of emotion through me, and my tears flow freely! During the intermission, I buy a leather wristband with the words "*Bring die Hoop weer terug*". *Another song to add to my Healing board*.

Back at home, I edit the song slightly and send *Pappa* a WhatsApp®-video of me reading the edited song:

> "There's a war inside me. It's for you, Dad, fighting cancer.
> Dad, I am telling you: you are good enough! So, let hope return!
> Dad, you are the light at the end of this dark tunnel.
> Dad, you are the beacon on the hill.
> I can see the man you are still going to be!
> So, let hope return!"[83]

> 13 August: Me: "This past weekend I saw Jo Black perform live for the first time in my life. Wow! What a phenomenal man! The Bruce Mason Theatre was packed, and he received a standing ovation."

I tell Hur about my edited version of "*Bring die hoop weer terug*". Maybe she would like to send her mom such a version?

Hur: "We were envious when we heard you attended '*Afrikaans is lekker*'! We had no idea the concert was happening last weekend and would have loved to attend! My son deeply appreciates Jo's raw music without any backtracks. Yes, please forward me the lyrics."

I scroll down the WhatsApp®-thread between Pappa and me until I find my edited version of the lyrics. I forward it to Hur and notice a new message notification.

Hur: "Oh, my goodness! How touching! As I was driving down Hukanui Road today, the song 'This is a move' started playing over the radio. Your dad's face appeared in my mind's eye. Once again, I realised that by faith, we can move mountains! We are in this together, united in Christ. Thank you for trusting me to support you on your journey. Mom said you can call her anytime with any questions you may have. PS: Remember to send me the lyrics! I am forwarding you a link to 'This is a move'."

Another notification comes through. With anticipation I click on the link. The first few words establish a tangible link between the Holy Spirit and my spirit. My eyes close as my hands raise in spontaneous worship. Every fibre of my being cries out to God as I sing along:

"Mountains are still being moved; strongholds are still being loosed!
God, we believe, 'cause yes, we can see that
wonders are still what You do!
Bodies are still being raised; giants are still being slain!
God, we believe, 'cause yes, we can see that
wonders are still what You do!"[84]

A fifth spiritual warfare song finds its way to my healing board.

*

Dr Caroline Leaf, a cognitive neuroscientist, specializing in cognitive and meta-cognitive neuropsychology, writes the following: "When we hope, it is an activity of the mind that changes the structure of our brain in a positive and normal direction."[85]

Hope!

Romans 15:13 reads as follows: "May the God of hope fill you with all joy and peace in believing, so that by the power of the Holy Spirit you may abound in hope."[86]

Our journey is blessed with so many testimonies and miracles. During one of my check-in calls with *Pappa*, he shared one of these miracles with me. He battled to urinate after the biopsy.

"Last night, around 3am, I thought I was going to die due to the unbearable pain! When I tried to urinate, the only liquid I managed to excrete was a tablespoon full of thick red blood. *Mamma* suggested we go to the EU, but I refused. It was too humiliating! Then she suggested we call our doctor-friend next door. But how can I call him at 3 am in the morning?

"Finally, Mamma said, 'Fine. Then we must pray!' Rosebud, I was in agonizing pain and not at all in the mood to pray. I told *Mamma* that she will have to do the praying then."

In our family, *Pappa* is always the one leading in prayer. *Mamma* is a strong believer, but not someone who will easily take the lead in prayer.

"My darling child, *Mamma* prayed such a beautiful prayer over me ... and half an hour later ..." I can hear *Pappa* swallow to bite back the tears. "Half an hour later, all my pain was gone, and the waterworks functioned the way it should!"

"Oh, *Pappa*! Praise the Lord!" I exclaim.

We wish each other a good night's rest and a blessed day, respectively. I fall down on my knees with my hands lifted high before a compassionate God. Such a holy moment between husband and wife, between God and them!

*

During this struggle, the Lord confirms again and again that Pappa's battle against cancer constitutes a spiritual war.

One Sunday morning, I was sitting in one of the pews, feeling a bit overwhelmed with spiritual fatigue. During our pastor's message, he reminds us that our fear cannot hinder God to get to us, since God meets us right where we are. The Lord confirms this with Scripture during my quiet time. The morning of His resurrection, when He appeared to Mary, all He needed to say for her to realize who He was, and that He had risen, was: "Mary!" She knew instantly that God was with her.

Unfortunately, the disciples didn't believe her. So, Jesus met them right there where they were, hiding behind locked doors. When poor Thomas refused to believe them, Jesus met Thomas right there where he was, stuck in his unbelief. Jesus lovingly invited Thomas to receive assurance the way he needed to: "Come, my precious child, come and touch Me! Come and feel that it is truly I who stand in front of you!" In Jeremiah 29, God promised: "You will seek Me and find Me when you seek Me with all your heart."[87]

*

14 August: I send *Pappa* a link to the website *Chris beat Cancer*. You probably recall that it "coincidently" appeared on my Facebook feed. I registered for the full program, which included a free buddy link. I forwarded the link to *Pappa*, which will enable him and *Mamma* to register for the program for free.

We are still awaiting *Pappa*'s test results. It is so frustrating! Apparently, the doctor is on leave and the results are waiting on his desk.

Meanwhile, Aaron's son is still battling whooping cough. Aaron sounds exhausted.

> Me: "Go look at yourself in the mirror and adjust your crown. Be brave! You are more valuable than many sparrows! God equipped us to overcome whatever stumbling blocks we may come across. It is

not easy. No matter how tough this journey gets, we still have so much to be thankful for. I am so thankful that you are my Aaron. If you weren't, people would have thought me completely crazy. Just imagine how many monologues I would have had to have between me, myself, and I if that were the case!"

Aaron: "Yesterday, one of the boys at school asked me about my son's health. Reluctantly, I admitted that it was not going all that well. '*Juffrou*, my dad and I prayed earnestly for him last night!' the boy exclaimed with such a sincere expression on his face. This morning, I prayed for eagle's wings. A few minutes later, I received a message from a friend about an eagle … We have Aarons and Hurs all around us, Sis!"

*

16 August 5:50 am: Me: "I can't breathe ..."

Aaron: "I know the feeling. But *Pappa* is looking good and full of faith."

I indulge in a nice long chat with *Pappa* and *Mamma*. We agree that they will visit during December. The moment I hang up, I phone our travel agent and book two plane tickets. Two years ago, I promised them that we would pay for their plane tickets when they turn seventy. My heart is overflowing with joy.

However, behind closed doors, Aaron is sobbing. She simply isn't ready to celebrate this Christmas without *Pappa*. Not after everything that happened this year. Because I love her more than life itself, my heart is breaking for her – even though she spent almost every Christmas of her life with *Pappa*, and even though she sees him almost every day of her life. My joy is resulting in her grief…

She apologises for dampening my joy. "No worries, Sis. Forgive me for stealing your chance to celebrate another Christmas with him." Immediately, Aaron regrets sharing her disappointment with me, but who understands her pain better than me, her soulmate?

Mamma oftens reminds us: "*Die mens wik, maar God beskik*[88]." God has already provided a solution for this predicament of ours.

*

The next few nights, Aaron doesn't sleep a wink since her son is coughing his lungs out. Sleepless nights slowly weaken her warrior-resistance.

18 August: The Voice from Eternity reminds Aaron that the enemy is trying to exhaust us, because he knows that he cannot beat us. Lovingly, the Voice reminds us to keep going, since the tides are turning. On this very day, Dylan Long encourages me to step out of the boat amidst the storm and to testify in advance regarding *Pappa*'s healing.

We receive an offer on our house the very next day. The tricky part is that they would like to move in on the 19th of September. Taking into account the planned two-week working holiday in Atlanta, we are left with only two weeks to pack for Atlanta and to pack up our whole house in preparation of the *Groot Trek*! One challenge after another awaits us.

20 August: We accept the offer to purchase.

A specialist diagnosis causes Aaron's son to be admitted to hospital. Her voice is almost inaudible due to the endless coughing in the background.

The Holy Spirit continues to remind us to praise and worship with songs of victory even before we receive confirmation of *Pappa*'s healing. He also reminds us to proclaim healing by saying it out loud.

26 August - Circle of Trust: Me: "*Julle!* I am so excited! The vendors to whom we made an offer to purchase, are currently attending a wedding in the UK. They accepted our offer in the midnight hour! Unfortunately, they can only vacate the property a week after we must vacate ours, but I am convinced it is a meagre obstacle to overcome.

"Everything is falling into place. Although we accepted a lower offer on our house in Hamilton than we would have liked, the vendors accepted our lower offer as well. In the end, we are saving more than we are losing.
"God provided again! Everything happened just at the right time – just as I said it would! I proclaimed: 'At the right time, our house will sell; at the right time, God will provide us with a new home in Auckland.
"And guess what? Aaron, do you remember that one weekend that we drove up to Auckland when I prophetically told you that I believe we are going to Auckland to have a look at our new house?"

My voice breaks as I am overcome with emotion.

"That house was one of the houses we viewed that weekend. Since it was way above our price range, we didn't give it another thought! Can you believe it? Praise precedes the victory![89]
"It always pays off to believe the promises God puts in our hearts. Aaron, I remember telling you: 'I am going to see my new house.' You asked me afterwards: 'So, what does the house that the Lord chose for you look like?'
"I can answer you now: Yes, I did indeed view my house that weekend – without even knowing it! God is simply amazing! To us it is the most beautiful house of all the houses we looked at.
"I will ask again whether the vendors can move out a week sooner. Regardless, I know the Lord will work all things together for good. I thank you all for praying with me. It has been a tough year. In hindsight, it has been a wonderful year! God is so faithful! Sending you lots of love!"

Me: "Hur, I'm forwarding my message to Aaron to you, so you can be updated as well."
Mamma: "Wow! *Baie geluk, julle*! We worship a mighty God!"
Other Sis: "I am so happy for you, Sis."

Aaron: "What a testimony! I am so happy for you too! We know just how long you were house hunting, and this is what we prayed for. I am truly happy for you. Please pray with me: my son is going back to school today for the first time in three months! He is very apprehensive because of his uncontrollable coughing."

Me: "I am praying calmness over your children and that they will feel comfortable among their friends. I also pray that your little girl won't be told off unnecessarily. (Aaron's daughter is facing a few challenges beyond her control. Not all teachers understand her situation).

"'I raise a hallelujah' has gone viral! I hear it everywhere I go! There is a universal spiritual war going on. Every day I am experiencing how God is turning this year into a victory year for us. We will raise our banners! All we need to do is to believe! Although we do not always understand why we need to wait, it is during the waiting time that God gets to work within us, and we get to grow in Him. God has a plan – a better plan than what the wisest among us can comprehend. Just cling to His promises. Believe it. Proclaim it."

Aaron: "My son and I have just shared holy communion together. He is so nervous about returning to school. My heart is breaking for him. We proclaimed our whole list of Scripture verses addressing fear. Thank you for interceding for us and for speaking life into the situation, Sis…" Her voice starts to break. "I really wish you were here!"

Me: "*Ag, ek is daar, man!* Just look in the mirror! I guess you didn't have a proper look this morning." I add a wink-emoji.

Meanwhile, Hur listened to my message and has just replied:

"My heart is thanking the Lord for the deep relationship between you and Aaron. I can hear how your spirits are resonating. I loved listening to your voice message!" *Ai,* Hur is such a sincere friend!
"I had to swallow a lump in my throat that just wouldn't stay down when you shared how you did indeed view your house that weekend and how it was above your price range. Our continuous chats on WhatsApp® these past few weeks have been so uplifting. I share in

your grief and in your joy. You always honour the Lord in all you do. You acknowledge Him in all your ways. I am overjoyed that the vendors accepted your offer and cannot wait to see your new house. I promise to come and visit!"

28 August: Aaron's voice sounds excited: "*Oe*, the Holy Spirit is just amazing! I shared my *skoonma's* testimony with my son this morning and explained how she was too scared to get out of bed after she broke her leg. Whenever she tried to, she would lose her balance, until she realized it was fear that crippled her.

"Hearing this, my son wrote all his fears on a piece of paper. I made a fire in the fireplace and explained to him that we are going to tear up all of his fears. 'This fire represents hell and we are going to throw all of your fear back into the pit of hell where it originated.'

"He ripped the paper into small pieces and lifted the grill. After throwing the pieces into the fire, he closed the grill, saying: '*Mamma*, now we are closing the gates of hell!' Today, we are once again attempting to return to school!"

School is indeed improving since the gates of hades have been closed. With every fibre in his ten-year-old body, Aaron's little boy is now fighting our spiritual war against *Pappa's* cancer with us. His young spirit is far older than ten. This little warrior also believes in *Oupa's* healing.

29 August: Circle of Trust: *Mamma*:

"To believe even though you cannot see it yet, is to work at it, to walk this road of the LORD, to refuse to get stuck in *cul de sacs*, to refuse to stare at walls, to refuse to quit the race of faith; to persevere like one seeing the unseeable God!"[90]

Hope.

"Therefore, since we have been justified through faith, we have peace with God through our Lord Jesus Christ, through whom we have gained access by faith into this grace in which we now stand. And we boast in the hope of the glory of God. Not only so, but we also glory in our sufferings, because we know that suffering produces perseverance; perseverance, character; and character, hope. And hope does not put us to shame, because God's love has been poured out into our hearts through the Holy Spirit, who has been given to us."[91]

Chapter 8
When the Rubber Hits the Road

> *Mamma*: "Many are the plans in a person's heart, but it is the LORD's purpose that prevails."[92]

The perfect house ended up being less than perfect. After a few worrying comments from the building inspector, we decided not to proceed with the purchase. I believe God is still in control and that He will help us to find the right house. Who knows, maybe we have already seen the right one as well?

Time for more house hunting before we fly out to Atlanta escapes us. We may be homeless soon, but we do have a brand-new caravan which we have only used twice. I phone Orewa Holiday Park and book a beachfront campsite for six weeks. A million-dollar-view for twenty dollars a day!

30 August: bad news flings us to the ground. *Pappa* saw the oncologist today.

> *Mamma*: "We saw two doctors today. Firstly, Doctor van Heerden administered a hormone injection. Then Doctor van Zyl discussed the oncology side with us. The cancer has five phases. *Pappa*'s cancer is in phase four. It has spread to his bladder, prostate, lymph glands and bone. Monday, *Pappa* will go for another scan in Witbank, which should detect any further spreading. Wednesday, he will meet again with the oncologist in Witbank. We believe that *Pa* will conquer this, and we praise God for all His mercy, goodness, and favour. Sleep well! *Lief julle almal baie*."

> Me: "By faith in His name, His name has made this man – Chris Viljoen - whom you see and know, strong. So the faith that comes through Him, has given him – Chris Viljoen – this perfect health in front of all of you."[93]
> *'Praise Precedes the victory!'*[94]. Jericho's walls fell because of their shouts of joy and not the other way around!"

My spirit is strong, but my flesh… Oh, dear!

I phone Hur.

Her first question is: "How is your dad?"

I take a deep breath before I answer her: "There are good news and bad news. The good news is that we are still standing firm and that we believe with every fibre of our being that *Pappa* will heal. The bad news …"

I am unable to continue. All Hur can hear are sobs. Hur does not need to hear anymore. She is all too familiar with these kinds of sobs – heart-wrenching sobs despite unwavering faith.

I do believe, Lord, but it is my dad!

Hur knows just how broken my heart is at this moment. She simply cries with me. We cry over my *pa* and her *ma*. We cry because we want to keep this cup from our parents. We cry because we know that God's will shall prevail and that we will have to bury our parents one day. We cry because we love them and want them with us – always. We cry because our emotions are so overwhelming that our bodies simply cannot contain them. Therefore, our bodies weep.

Hur knows my heartache does not mean that I stopped believing. She knows I am just crying because … it's my dad!

31 August: I cry the whole day. In between all the tears, I pack. Every now and then, I sink to the ground and weep. The last time I missed Aaron this much, was the day the gate fell on Dad's head. I declare anew that we are locking the gates of hell. Aaron sends a message, confirming that she also cannot stop crying.

Lord, it's our dad! We raise a hallelujah louder than the unbelief![95]

By the way, what my dad didn't tell me on 30 August, and what I only became aware of much later, was that the oncologist only gave him eighteen months to live – if he undergoes chemotherapy! *Pappa* replied: "Thank you, but I can die much cheaper!"

That is why one should choose your Aaron and Hur carefully. If your Aaron believes in the same goal as you do but in a different way to get there, it may not be very helpful. It would have been catastrophic to allow people into our inner circle who believed chemo was the way to go, regardless of the strength of their faith. The chemo expert himself did not even believe that chemo would heal my dad. If your Aaron or Hur does not agree with both your end goal and your plan of action, they will desert you when the night is darkest.

1 September: Aaron: "At this stage, I am sharing my journey with you alone."

3 September: Me: "When our whole being cries 'Glory to God' when all our thoughts are focused and centred on God, when we meditate on Him, then there, where we are – it may be in an automobile, it may be in a kitchen – but it becomes God's temple because everything is saying: 'Glory!'"[96]

Ma: "The best defence against the (enemy) is to be so 'God-centred' that we give no place to him (the enemy)."[97]

4 September, shortly before we need to leave for the airport, I receive the Whole-Body PSMA Scintigraphy report from *Mamma* with a footnote: "Really bad." According to the report, the cancer has spread to *Pappa*'s pelvis, lower back, femur, and ribs.

I take one look at the report, print it out, write "Healed in Jesus' Name" in big, fat, red letters over it and send it back to my parents. In the bottom corner of my healing board, I glue a red envelope with the opening to the front. The report with the damning diagnosis gets folded into a small square and placed into the envelope. My purpose is to keep it as a future reminder of how great our miracle will be.

Aaron stared at the scan with several dark spots, indicating cancer, and said: "At least there is a lot more white than black."

We still believe without a doubt that *Pappa* will be healed, even though the scans and reports briefly took the wind out of our sails.

7.23pm: Me: "We are boarding. *Liefde, liefde*."

My sails are still gasping for some wind.

During the flight, I watch the film "Breakthrough". It serves as confirmation that *Pappa* will be healed. While waiting at Chicago Airport for the boarding announcement for our connecting flight to Atlanta, I send a message to my parents and sisters, urging them to watch the movie. Aaron replies almost immediately. She told *Pappa* and *Mamma* about this movie last night but couldn't remember the title. Considering the time difference, it was probably at the same time I was watching it... "The spirits of prophets are subject to prophets."[98]

Every time I experience a revelation from the Holy Spirit which resonates with what Aaron or Hur, my prayer partners, are sensing in their spirits, reassurance covers my heart like a soft down blanket. Each time, it confirms that I am not merely hearing what I want to hear. The Holy Spirit whispers in our ears, and the sound of His Voice resonates across oceanic and air currents with our spirits.

13 September: *Ma*: "Positive thoughts lighten the weight."

14 September: Mamma sends us a photo on the Circle of Trust-WhatsApp® group chat of an eagle protecting her nest. It looks as if the nest is surrounded by flames. When you turn the photo upside down, it looks like Jesus' face, with a crown of thorns on his head, in the centre of the flames.

15 September: Me: "Now faith is the assurance of things hoped for, the conviction of things not seen."[99]

No, we are not being naive. Neither do we live in Lalaland. We know we are in the middle of a raging fire. We also know Who our Protective Eagle is, amid this fire!

*

Monday, 16 September 4:48 am, we touch down in Auckland. We only have three days to pack up our Hamilton home because the removal company arrives on Thursday. I change gears to autopilot.

*

For various reasons, our family was not in favour of chemo. *Pappa* finally agreed to radiation therapy, but we didn't have peace about it.

16 September: The medical insurance rejects a pre-approval for radiation therapy. The oncologist writes another clinical recommendation, which the medical aid rejects a second time. All of us were at peace with this decision. We were not convinced that radiation was part of God's healing plan for *Pappa* anyway.

Please take note, dear Reader, that this is our journey, our testimony, our healing journey. I am not advising you to reject chemo- or radiation therapy if you have cancer. All I am saying is that we knew it was not God's will that we take that route.

When he received the initial diagnosis, my dad had already decided to use alternative healing remedies. He trusted the guidance of the Holy Spirit in this regard, and above all – He relied on God for his healing. The remedies he used included Cancer bush, Essies tea and a product which stimulates the production of new stem cells. Chris Wark's advice also came in handy. Chris focuses mainly on a changed lifestyle, which includes a cancer-fighting diet packed with fruit, vegetables, and herbs. Eating food that strengthens your immunity and avoiding cancer-feeding food, is key. But Chris' greatest weapon against cancer was his faith in Jesus Christ.

Aaron and I help *Mamma* with ideas to incorporate some of Chris' suggestions into *Pappa*'s daily diet. My parents decide to do that which is possible and sensible to them. Like it was for Chris, our greatest weapon is our faith.

> 17 September: Other Sis: "The giant in front of you is never bigger than the God living inside of you."[100]

18 September: We all sing "His truth is marching on" and declare it over *Pappa*.

Early morning, 19 September, two large removal trucks pull into our driveway.

Ma: "*Tannie* Mini, *Tannie* Kiewiet and Monique are coming to visit on Saturday. Sanel and Hendrik will also be here. Please let me know if you can bring a dish to share."

Other Sis: "I will make a pasta dish for Saturday and a potato dish for Sunday."

Ma: "*Baie dankie, my kind.* I appreciate it. Oh, something for tea on Saturday as well. Peppermint tart and maybe a Swiss roll?"

Other Sis: "I will bring a Swiss roll."

Lin: "I will make pumpkin pie."

Ma: "*Dankie,* Lin. I appreciate it. We all love that pie."

Me: "*Sjoe*, it sounds delicious! I will visit in the spirit. My spirit will bring *appel-lief-julle-tert* with *seën-sous.* It is non-fattening! LOL. Please think of me! We are moving today."

Ma: "Good luck with the move, *my kind.* Are you moving into a rental?"

I roll my eyes, and while supervising the move, I send a quick, out-of-breath voice note:

"No, *Mamma.* Can you remember I told you that we are going to camp next to the beach until we find a new house? Love you all. Talk later."

Ma: "Of course, now I remember. *Ai toggie*, where has the memory gone!"

Poor Mamma. She barely has a moment's rest these days.

Lin: "Thinking of you, Sis."

Other Sis: "Good night."

Early Friday morning, SA-time: *Ma*: "Is the move done, and have you put up camp?"

I send a voice note:

"Hello, *Mamma*. Thank you for checking in. Niki and I have arrived in Orewa. We booked into a motel across the street from the Holiday Park. Hubby is still on his way from Hamilton, towing the caravan. He waited until they finished loading and handed the keys to the new owners.

"He is reluctant to leave the caravan alone overnight. So, he will be sleeping there tonight. I have no idea how he will manage to climb over all the baggage to get into bed, so good luck to him! Niki and I prefer to sleep in the motel tonight. We will sort out the campsite tomorrow morning. Please wish us luck!" I giggle nervously. "You know that setting up camp can be a bit of a test to one's relationship. Between you and me – we had to pass more than enough tests lately, *maar nou ja*. Niki and I are having a rest while watching Showmax. We'll take a stroll to one of the beach cafes down the road later ..."

"A Jazz-place," Niki interrupts.

"*Ag*, it is such a cute little restaurant," I continue.

"It is … what ... *Down* …?" I ask Niki.

"Downtown … Downbeat!" Niki remembers.

"*Ja*, Downbeat. They play old-fashioned black-and-white silent movies. Niki and I love the ambience there. Well, we are now officially homeless – with lots of money in our bank account," I laugh.

"Now we just need to find a new house in Auckland – and then we'll have no money left in our bank account!" I jokingly add.
"Please think about us, pray for us. I believe the dear Lord Jesus has already chosen the perfect little place for us. In the meantime, we are preparing for our camping adventure. We are going to camp on the beachfront for 'an indefinite period'. Oh, please pray for the weather! September here, is nothing like September over there by you guys. But we were born in Africa. We are tough as and have endured Kiwiland's wet and cold for eleven years now."

"We can make it, eh, Niki?" I try and include Niki again in my one-sided conversation.

"You have been speaking for two minutes and forty seconds. Too much! Stop!" Niki complains.

"Okay, Niki said I have to stop," I conclude with a chuckle. "I love you so much!"

"Niki, you cannot stop your *ma* just as she is getting warmed up!" Lin pretends to be outraged. "You must understand, our heart's desire is to talk to one another and share our exploits. So, you need to show a little more understanding for your *ma*. *Siestog*!
"*Ag*, my dear homeless family, honestly, we can offer you a place to sleep here in South Africa. We know it's hot – thirty degrees Celsius and the children have no choice but to swim every day. We shall take pity on you and offer you a place to stay because we truly love you!" Lin jokes rather earnestly.
She chuckles and adds: "I shall pray for pleasant weather. Enjoy your time there – and especially this evening's dinner. That restaurant sounds cute. I would love to hang out there with you. Love you so much."
Me: "*Ag*, *dankie*, Sis! I'll take you up on that one. Niki and I will hop onto the Concorde, and we'll be there within seconds!" I jokingly daydream along. "Then we can have a warm night's rest, you can spoil us rotten, and we can have a *lekker kuier* on Sunday. That will be fantastic!"

Niki shakes her head and laughs at my silliness. But underneath all the jokes and chuckles, a few tears glisten on either side of the great ocean...

*

Just after five the next morning. Niki and I wake up. I switch on my phone, and *Mamma*'s voice note vibrates:

"Is your campsite all sorted?"

"*Nee, Mamma*," I answer in my hoarse wake-up voice. "Settlement and registration were finalised at 4 pm yesterday afternoon. New Zealand solicitors strongly advise their clients against handing over the keys before registration has been confirmed, so hubby arrived when it was already dark."

I interrupt myself to take a sip of hot coffee.

"Our poor cats are only allowed to camp with us for three days," I continue. "Campground rules do not allow pets for a longer period. Currently, our only option is the cattery down the road, which works out rather expensive. Please pray for alternative accommodation for our feline children?"

My throat burns as I take a huge sip of coffee too quickly.

"Wish us luck for today. Big camp erecting day! Since we do not know for how long we will be nomads before we manage to buy a house, I needed to bring a lot of stuff along. Niki's Homeschool books, winter- and summer clothing, *ag*! So much to sort out! Well, we are now officially entering a working holiday: Hubby drives to the office every morning with great relief. Instead of spending almost two-and-half to three hours on the road, it takes him only two-and-half to three minutes to get to work!

"As soon as he leaves, Niki and I will sort out our campsite. As of tomorrow, it will be back to school again – in the caravan! House hunting will be done in the afternoons and over weekends. Luckily, my household chores will be slightly less than before! Have a nice day!"

*

Exhausted but thankful, I climb into bed. I sleep like a baby. Early the following day, I watch the sunrise through the window while warmly tucked in underneath my cozy wool duvet. Right in front of our campsite a footpath meanders parallel to the beach. The ocean glistens with a shiny blue tint while a few gentle waves brake quietly onto the sand. Shakespeare Park is basking in the sun on the horizon at the opposite end of the bay. A palm tree in the corner of the caravan window completes the picture.

20 September: The rubber hits the road *lekker* hard as my *hartkind* wins the 200-metre hurdles, boys U/16 at Twizza[101]. Tears role as I watch the video. Two weeks later, he finds himself in hospital, diagnosed with Guillain Barr – an illness attacking your nervous system. He is unable to lift his legs…

21 September is the big *kuier* with the *appel-lief-julle-tert* and family from all over the country. For a second, my thoughts run haywire: *This time they really are coming to say goodbye!* I am heartbroken for not being able to join them. If only I could do more for *Pappa*!

23 September: I receive a phone call from *Mamma* during dinner. An ice-cold iron fist clenches around my heart. *Mamma* sounds tensed, and her words sound disorderly: "*My kind, Pappa* doesn't know how to tell you this … He won't be able to fly."

Klingel, kletter, tingeling … Duisend splinters om my heen…[102]

Chapter 9
The Armour

The most well-known instructions regarding the Christian's armour are possibly found in Ephesians 6, where Paul heeds us to put on the full armour of God. Why? He explains as follows:

> "Take everything the Master has set out for you, well-made weapons of the best materials. And put them to use so you will be able to stand up to everything the devil throws your way. This is no afternoon athletic contest that we'll walk away from and forget about in a couple of hours."[103]

In Daniel, chapter 10, we read about a vision Daniel had, which drained all his energy. It was so bad that he prayed and wept for three weeks. Then an angel appeared to him:

> "Suddenly, a hand touched me and raised me to my hands and knees. He said to me, "Daniel, you are a man treasured by God ... Stand on your feet, for I have now been sent to you ...
> "Don't be afraid, Daniel ... for from the first day that you purposed to understand and to humble yourself before your God, your prayers were heard. I have come because of your prayers. But the prince of the kingdom of Persia opposed me for 21 days. Then Michael, one of the chief princes, came to help me after I had been left there with the kings of Persia ..."

Daniel explains that his strength returned when the angel touched him:

> "Don't be afraid, you who are treasured by God. Peace to you; be very strong!"
> He said, "Do you know why I've come to you? I must return at once to fight against the prince of Persia, and when I leave, the prince of Greece will come. No one has the courage to support me against

them except Michael, your prince. However, I will tell you what is recorded in the book of truth."[104]

In Zechariah 4, Zechariah, too, had a vision. An angel gave him a message to give to Zerubbabel, who governed Judea during that time:

"This is the word of the Lord to Zerubbabel: 'Not by strength nor by might, but by My Spirit,' says the LORD of Hosts."[105]

Paul called the Corinthians to spiritual warfare, as well:

"The world is unprincipled. It's dog-eat-dog out there! The world doesn't fight fair. But we don't live or fight our battles that way - never have and never will.
The tools of our trade aren't for marketing or manipulation, but they are for demolishing that entire massively corrupt culture.
We use our powerful God-tools for smashing warped philosophies, tearing down barriers erected against the truth of God, fitting every loose thought and emotion and impulse into the structure of life shaped by Christ."[106]

Tonilee Adamson and Bobbye Brooks explain spiritual warfare as follows:

"As Christians, we are in a spiritual battle of some sort on a daily basis. In warfare, battles are fought on different fronts, for different reasons, and with varying degrees of intensity. The same is true in spiritual warfare. Our spiritual battles and warfare are real, even though we cannot physically see the attacker. But we can educate ourselves on how the battles are fought and how they impact our lives on a daily basis."[107]

Focusing on 'the sword of the spirit' – the 'Word of God' – I study Ephesians 6 again, where Paul heeds us to put on the full armour of God.

I already know this is a spiritual war. A picture of King Arthur's sword, Excalibur, in my one hand and a big, heavy wooden shield in my other hand, appears in my mind's eye. I need faith – a shield "to extinguish the flaming arrows"[108] of the enemy – and God's Word – a sword – to attack my adversary.

While pondering on the huge, traditional wooden shield, my thoughts jump to my favourite super-hero movie, Avengers, and Captain America's smaller, rounder, red and blue shield. A bit of Marvel research convinces me that Captain America's shield is the more appropriate armour here since he does not only use it as a weapon of defence, but it is also his primary weapon of attack:

> "Captain America relied heavily on the shield as a weapon that was balanced in both defensive and offensive capabilities. Made from vibranium, it is able to absorb and reflect kinetic energy enabling it to withstand impacts that would break a shield made of any other material."[109]

The Voice from Eternity concurs: "Your faith should be so strong that no attack can penetrate it!"

God's Word to attack with, my faith as a shield to extinguish enemy attacks, and my healing board as my war strategy. Slowly, step by step, I revise my war strategy.

I realise anew that I am powerless in my own strength. In humble dependence – like Paul of old – I stand in front of the Throne of grace and plead: "I claim to know nothing – except Jesus Christ and Him crucified!"[110]

"Cancer, you cannot have my dad! Because CHRIST WILL HEAL my dad! Hallelujah! Praise precedes the victory![111] Jericho's walls fell because of their shouts of joy and not the other way around," I confirm the war cries on my healing board.

Aaron, Hur and I. Together we pray for a miracle like never before "because when God's warriors go down on their knees, the battle is NOT over. It has just begun!"[112]

With determination and emphasis, I read every Scripture verse out loud and declare with fresh commitment:

"For we wrestle not against flesh and blood, but against principalities, against powers, against the rulers of the darkness of this world, against spiritual wickedness in high places."[113]

"But faith takes us beyond ourselves to where God is".[114]

"By faith in His name, His name has made this man – Chris Viljoen – whom you see and know, strong. So, the faith that comes through Him has given him – Chris Viljoen – this perfect health in front of all of you … "Let it be known to all of you ... that by the name of Jesus Christ the Nazarene – whom you crucified and whom God raised from the dead – by Him this man is standing here before you healthy."[115]

The same Power that raised Christ from the dead is living in me![116]

"'Lord, if you are willing, You can make me clean.' Reaching out His Hand He touched him, saying: 'I am willing; be made clean.' Immediately his disease was healed."[117]

"Look, I have given you the authority to trample on snakes and scorpions … nothing will ever harm you!"[118]

"Therefore, I declare in the Name of Jesus Christ: Cancer, you cannot have my dad! CHRIST WILL HEAL my dad!"

Full of determination, my fingers start typing:

"*Pappa*, the doctor's prognosis is but that of a human being. It is not how it has been written in the heavens! I cancel this prognosis in Jesus' Name and speak life over you: healing and not illness. Thank you, Holy Spirit, that You are interceding for us with groans that words cannot express since we do not always know what to pray."[119]

With a loud voice, I start singing each of my war songs while their printed lyrics are dancing to the tune in the light morning breeze, dangling from their silver strings. The devil can really mess with one's head when he is in the mood. What chases him away the quickest, is music. God's music!

> "Whom shall I fear – I know who goes before me, I know who stands behind. The God of angel armies is always by my side … Nothing formed against me shall stand. You hold the whole world in your hand. I'm holding on to your promises."[120]

> "I raise a hallelujah, in the presence of my enemies. I raise a hallelujah, louder than the unbelief. I raise a hallelujah; my weapon is a melody. I raise a hallelujah; heaven comes to fight for me. I'm gonna sing in the middle of the storm. Louder and louder, you're gonna hear my praises roar. Up from the ashes, hope will arise. Death is defeated; the King is alive!"[121]

> "He made me a promise. He gave me a calling; I know that He's able to keep me from falling. I focus my hope. The past is gone. The moment to prove the power of God has come! In this I will be confident! The time is now, the odds are long, the haters are loud, the pressure is on, but devil, you messed with the wrong one. I got my sling, I got my stone, and in the name of my God I will overcome. In this I will be confident."[122]

> "Mountains are still being moved; strongholds are still being loosed. God, we believe, yes, we can see that wonders are still what You do. Bodies are still being raised. Giants are still being slain. God, we believe, yes, we can see that wonders are still what You do."[123]

I scroll down our WhatsApp® thread until I reach Aaron's message:

> "A warrior is that woman who gets up despite the enemy trying to destroy her. A woman who declares victory before seeing it. A woman

> who believes she will receive her miracles, because she knows the Lord she serves is alive and powerful."[124]

This warrior is ready for battle. Some days are good; some days not so much. On the good days, I proclaim our testimony, encourage Aaron and Hur and provide support to *Pappa* and *Mamma* with phone calls, messages, silly banter and a spice of humour wherever I can.

Other days hit me hard. On these days, I hide in my study behind closed doors, turn up the volume of my war music and grab my dad's healing board. To the background of my warfare music, I read out loud everything written and pasted on the healing board until my deafening war cries chase the enemy away. My faith destroys him with every song and every piece of Scripture.

Sometimes, when it feels as if the flames of this war are scorching the hair off my body, I go lie flat on my face before the Throne of Grace. With my last bit of human strength, I cling to the seam of His garment with both hands. During times like these, all I can utter is 'Andante, Andante …'

> "In addition to all this, take up the shield of faith, with which you can extinguish all the flaming arrows of the evil one. Take the helmet of salvation and the sword of the Spirit, which is the Word of God."[125]

If the enemy comes again – despite all my warfare – and whispers softly over my shoulder: "What if …" I turn around. I stare him square in the face and tell him: "I do not fear this storm! I will step out in faith, and should the water overwhelm me, I know He will save me. I'll much rather be soaking wet and in the arms of Jesus than safe and dry in the boat. Thank you very much for your concern. But God's got my back!"

*

There were times when the devil attacked me viciously – especially during August. Years ago, a pastor's wife relayed how the devil attacked her during her time of spiritual warfare. Sometimes she would see him standing in the hallway. After a while, she got so bored with his unwelcome appearance that she simply said: "Oh, you, again!" This caused him to lose courage and disappear.

Initially, I thought something like that would scare me to death. The more I thought about it, the more my spirit revolted. I declared: "No, Lord! I refuse to fear a little demon." I asked God to help me to rule over the devil, should he ever appear to me.

One August night, I had a terrible nightmare. When I woke up, I saw a little man – actually, more the shadow of a little man – doing a Rumpelstiltskin-dance in front of my bed. It was as if he was trying to cast a devilish spell over my dreams. I sat up and said out loud: "In the Name of the Lord Jesus: *wyk, satan*!" Instantly, he disintegrated and disappeared. To top it all, I knew no fear! I could immediately sense that the evil spirit had left, and I was able to fall asleep straight away.

Alas, he did not give up that easily. The following night, I woke up again and saw him standing in the doorway of my bedroom. Once again, I sat up and said: *"Wyk, satan!"* This time, the figure remained. First, I thought it was because I forgot to add: "In the Name of the Lord Jesus," but then the figure became clearer.

I saw the outlines of an angel as if a bright light was shining from behind it, filling up the full length of the doorway. In front of the angel was the odd, shadowy little man. Then the angel lifted his sword above his head and sliced right through the *tokkelossie*-figure! The little man disintegrated instantly.

The moment I realised it was an angel standing in the doorway, the image disappeared. Intense peace filled me.

But the cat with nine lives (or in this instance, the devil and his henchmen) returned a third time. When I woke up this time, they appeared in the form of three shadowy, big-headed, one-eyed snakes, slithering up and down my doorposts. I sat up and, rather irritated, said: "Really, now? You know by this time that you are unable to enter my room. In the Name of the Lord Jesus: *wyk, satan*!"

After this, they stayed away.

It was interesting how the enemy noticeably fled further and further away from me. Initially, he stood in front of my bed by my feet; then he stood in the doorway to my room and finally, he sent three pathetic little henchmen who could do nothing more than to slither up and down the door posts.

When I told Niki about this, she said: "Mom, you should anoint your door with oil." Without avail, I got up and anointed every single doorpost in my house with oil.

"Mom, I didn't mean every door in the house!" Niki sighed, rolling her eyes.

The enemy didn't dare to come near my house again.

> "Resist the devil, and he will flee from you."[126]

It isn't easy, though! Hur and I are both shattered by the crushing scan reports of our respective parents. Our faith remains unwavering, but our hearts are broken, because they are our parents!

Here we are: two daughters who believe with every fibre of their being that their respective parents will be healed. At the same time, medical science is condemning her mum and my dad with words like "Phase Four" – as if they have power over life and death!

Still, we raise our hallelujahs up high in the presence of our enemies and louder than the unbelief.[127] The worse the news, the more we believe in the healing and the testimonies we proclaim.

My phone beeps with a message from Aunty Joyce:

> "Trusting God allows us to enter His rest, and rest is a place of peace where we are able to enjoy our lives while being confident God is fighting our battles."[128]

Chapter 10
Then Our Tent Blew Away

Mamma explains that *Pappa* is too ill to fly. He is unable to sit down for longer than half an hour. He battles to sleep and struggles with all sorts of symptoms. One thing I am blissfully unaware of (and which I would only find out months later) – is that *Pappa* told *Mamma* that she needs to prepare herself …

Blinded by tears, I stare at the splinters at my feet. Two years of dreaming and planning their Christmas holiday with *me*; all the healthy, delicious recipes I was going to spoil *Pappa* with – shattered! They are different when they are visiting me here in New Zealand. They tend to forget a bit about all the other children and grand-children and all the challenges of the workshop. Here, I have their undivided attention. Here, Niki also gets a chance to get to know her *Oupa* and *Ouma*, as other kids do …

Hubby shares his practical solution to my predicament: "Do what you need to do. I won't be able to fly to South Africa again this year since I need to fly out there for business in February. It is too close to year's end to be away from the office right now, anyway. Cancel *Pa* and *Ma*'s flights and book tickets for you and Niki."

Nou ja, I sigh. Just like *Pa* taught me to do from an early age, I take a deep breath, wipe my tears, and straighten my shoulders. My voice is still quivering, but I manage to avoid a teary outburst during my brief telephone conversation with our travel agent. She cancels *Pa* and *Ma*'s bookings. We lose a great deal of money in the process, but that is the least of our problems at this stage. My only worry is to get to South Africa in time. *For everything else, there's Mastercard!*[129]

I decide to surprise *Pappa* on his seventieth birthday. *Mamma* and Lin keep our arrival date a secret. We tell *Pappa* my little girl and I will be there for Christmas. With that, we once again lay our large, heavy burden at the feet of Jesus and put the sadness behind us. Aaron, Hur and I focus on our spiritual warfare.

Every now and then, the enemy hovers over my shoulder and whispers: "Do you really think you are going to arrive there in time?" Like Steven Furtick the Second, I push him forcefully to the ground, turn around on the dime and proclaim with a steady conviction:

> "I will be confident in this: He made me a promise! He gave me a calling; I know that He's able to keep me from falling! I focus my hope! The past is gone! The moment to prove the power of God has come! In this I WILL be confident!"[130]

Fortunately, the Lord fills up my daily duties with a myriad of distractions. Caravan life has its own challenges – especially in Auckland during September and October! A cyclone over the Pacific causes severe rainfall and wind.

Meanwhile, life continues: my other nephews and nieces win athletic races, *Mamma* turns seventy, and we are still camping on the beach. We homeschool, do house hunting, and tend to our washing at the local laundromat.

Amid all of this, *Pappa*'s health continues to deteriorate.

One morning, we view a property which had come up repeatedly during our internet searches. The asking price was way above our budget. Honestly, we do not want a mortgage again!

However, the house on the hill has a two-bedroom minor dwelling which can be rented out to cover the mortgage. The jump between a property just big enough to house the three of us and all our animals, to one which includes a second dwelling, is a couple of grand more than we wished to spend. On the other hand, it is a struggle to find a property that ticks all our boxes.

We jump through a few hoops, view a few more properties in the area and reluctantly consider the idea of another mortgage. The view, the openness, and the proximity to hubby's office, move the property to number one on our list. We present our final offer on October 1st, with a few amendments to the original one. By the sounds of it, we'll be able to move in just before Christmas.

The end is nigh!

*

8 October: an ambulance transports my *hartkind* to Pretoria, where he will undergo further neurological testing.

9 October: a neurologist confirms that there is no sign of nerve damage and that my nephew will recover fully, despite a long and arduous journey ahead of him.

11 October: he gets released.

Hallelujah!

*

After four weeks of beachfront camping, we are literally blown away on October 16th! Niki and I are busy paying for our groceries at Countdown when I receive a call from hubby. He is leaving the office as we speak and urges us to get to the campground immediately. Camp management phoned him and informed him that the wind has blown our tent right over the caravan.

While scrambling to load the groceries into the SUV, my mind rushes to another windstorm on the beach…

*

I am eight years old, other sis is five and Lin is a wee baby in a pram. The five of us are happily camping at the caravan park in Richards Bay. This morning, we threw down our towels on the right side of the *dolosse*[131] at Alkantstrand. (That is the right side, if you are facing the sea.) We play joyfully in the waves.

By lunchtime, we all feel rather hungry, sunburned, and exhausted. *Pappa* and *Mamma* decide it is time to return to our campsite. We have everything packed up in a jiffy and proceed to the other side of the *dolosse*. *Mamma* pushes the pram, *Pappa* carries the chairs and cooler box, while other sis and I carry our towels and the plastic buckets and shovels we used to build sandcastles with.

Everything proceeds smoothly and orderly until we reach the other side of the *dolosse*.

A horrendous wind almost blows us over. Other sis and I complain that the sand is hurting our eyes and legs. *Mamma* suggests we cover ourselves with the towels to protect us against the stinging sand. The wind is fierce. I am unable to look up ahead of me and battle against the wind through the thick, soft sand. Fortunately, the pram is leaving deep tracks in the sand. Just like Hansel and Gretel, I simply look down and follow the tracks.

At one stage, I hear *Mamma* calling out other sis' name. I turn around to look for her and notice that she is walking towards the breaking waves. The strong wind pushes her little body askew and forces her closer and closer to the ocean. *Pappa* puts down everything he is carrying and runs to fetch her.

"Just look down and follow the pram's tracks, Sis," I present my logical solution.

When we reach our campsite, I feel great pity for my poor legs covered in hundreds of little red dots where the sand hit them. *Mamma* carefully removes the baby blanket she used to cover the opening of the pram to protect the baby against the sand. "Oh, my goodness! Everything is covered in sand!" she exclaims. Lin slept peacefully all through the sandstorm, even though her ears were filled with fine sand. *Mamma* even shook sand out of her nappy when she took it off. Sand dust simply invaded the pram.

Even though other sis was almost blown into the sea, fear never crossed my mind. As long as I followed the pram's tracks, I knew that Pappa and Mamma would lead us safely back to our campsite.

*

As we enter the Orewa Holiday Park, the windstorm at Alkantstrand flashes before my eyes. Our belongings are quickly being blown towards the sea – just like other sis years ago. The brand-new awning with its poles and pegs, have been blown over the caravan and are dangling on the other side of it. We run around like mad chickens trying to save our precious belongings from a sure sea death. Never before had these three Standers succeeded in working together so effectively. We packed up camp in record time.

My healing board was safely stacked on top of the portable camp closet. Niki discovers it where it is tucked behind the closet. Miraculously, the wind did not blow it away. What was missing, though, was one of the spiritual warfare songs – "I raise a hallelujah".

Ag wat, somebody who needs to read it, will probably pick it up, I think to myself. *Maybe it can add to a miracle in that person's life. I can simply print the lyrics again and attach a new copy to my healing board. That is, of course, whenever I can unpack my printer again!*

Supernatural protection prevented the healing board from being damaged. A very interesting occurrence is that the MRI results, which were safely tucked inside the red envelope, blew out of the envelope. Immediately, I accepted it as a sign that those results were irrelevant. Later, I would think back on this day – 16 October – and smile about the fact that both the MRI results and this specific song blew away on this particular day.

Two hours later, our two vehicles are packed to the brim with all the contents of our former tent, including the tent itself, a few broken tentpoles, and hopefully most of the tent pegs. Hubby and I give each other one look and exchange similar thoughts. "I think we are done camping," we say simultaneously.

We are in luck. The motel across the street where Niki and I spent our first night of 'homelessness' has a two-bedroom apartment with a bathroom and kitchenette available. We urgently start looking for a furnished, short-term rental where cats are allowed, for this exercise is becoming very costly!

*

20 October: my *hartkind* contracts tick fever.

22 October: the Lord sends me a message through Joel Osteen, introducing a different perspective. He puts it this way: David never defeated Saul – he survived him. He had to resist him repeatedly, and one day, when he glanced over his shoulder, Saul was gone.

Some enemies are not ours to defeat; they are there to build our character. If we can resist the enemy, he will flee from us, just like James promised in chapter four verse seven.

Meanwhile, my mum, baby sister, and I are ardently planning my parents' seventieth birthday party, which I will not attend (as far as everyone else is concerned). *Mamma* asks me again to write a birthday invite. *Pappa* and the rest of my family are still unaware of my part in planning this party. At least the excitement of the surprise helps to keep us positive – that – and the spiritual war we wage every day.

Our new congregation (or I should rather say, our old one – the first congregation we joined upon our arrival in Kiwiland) is hosting a ladies camp the first weekend in November – the "Hope" Ladies Camp. I see it as a shining beacon on my journey. The Holy Spirit urges me to go. Sunday morning, I write down Hur's name, her mum's and her sister's names, as well as my own name on the camp registration list before anyone else has time to register.

Isn't God good to me? Amid my confusing life, He brings me and my fellow spiritual warriors to a camp themed, "Hope" – situated next to one of the most beautiful beaches in New Zealand!

Throughout the weekend, several ladies, who lost one or both of their parents, testify. With every testimony, the enemy grips my heart with an ice-cold, steel fist. Yet, I resist him! Another young woman shares her raw heart with us on Sunday morning. She explains how a friend of hers flew to South Africa to say goodbye to her dying mum. When she arrived at the airport, her family had to inform her that she was too late…

I grab Hur's arm and start to cry, "Please, Lord, let me be in time! The warrior inside of me arises through the tears. Lord, I refuse to accept defeat!

"Hur," I whisper through my tears, "I refuse to give in to the fear that I won't be in time to embrace my dad!"

Hur straightens and turns to face me. She locks eyes with me and prophesies: "Lijlanie, you are going to South Africa to celebrate your dad's healing!"

The guest speakers are done, and we close with praise and worship. Cathy leads us in worship while her mother, Petro, accompanies her on the piano. They conclude with none other than, "I raise a hallelujah". At once, I know: God planned it this way! At the end of the song, I get up, approach Cathy, and ask if we could sing it once more. "May I please borrow your microphone for a few minutes and explain to the ladies why I want us to repeat the song?"

"Of course," she smiles.

I turn around to explain my request to the group of ladies. I was determined not to cry in front of them. Seeing so many women with raw emotion all over their faces just got to me! Clearly, everyone has been deeply touched by the morning's testimonies and worship. As the words tear through my body, the tears start streaming.

"My heart goes out to every single one of you who lost a parent… But, today, there are three daughters among you who are fighting for their parents: Hur and her sister are fighting for their mum, and I am fighting for my dad. We refuse to lose our parents to cancer! 'I raise a hallelujah' is our war song. Will you please repeat it with us so we can all intercede for our parents?"

I hand back the microphone to Cathy and return to my seat next to Hur. Petro starts playing the intro to the song. This time, it sounds as if twice as many women are worshipping. Every woman gathered here, raise a hallelujah from the bottom of her heart! Overcome with emotion, I go down on my knees. With my hands lifted high, I pour out my soul in front of Him. *Lord, it is my pa! It is Hur's ma – mý New Zealand ma!*

When I say goodbye to my fellow campers, I look into their tearful eyes as I receive encouraging hugs and sincere well-wishes. The camp ends as the ladies start leaving one by one. When I turn on the ignition, an intense longing stirs up inside me, and I phone Aaron. It doesn't ring long before she answers.

I share the morning's worship time with her. Halfway through, Aaron interrupts me: "Sis, didn't *Pa* and *Ma* share the news with you yet?" She sounds half surprised, half guilty.

"What news?" I ask, confused.

"The last bio-plasmic scan was clear!" *Pappa* underwent bio-plasma scans, which identified blockages within his body. These blockages can be neutralised through electrodes. The machine can identify cancerous cells and specifically target them.

"The scan showed that the cancer in the lymph, bladder and bone were cleared. There was only a small amount present in his prostate gland," Aaron explains.

I am overcome with joy. Together, we celebrate God's faithfulness. "When did you get the results?" I ask.

"About two weeks ago. I am so sorry, Sis! I honestly thought *Pappa* and *Mamma* told you. Things were so crazy over here ..." she tries to apologise.

"Wait, I have to phone *Pappa* and *Mamma* immediately," I cut the conversation short.

Pappa confirms the good news. "*Ai*, why didn't you phone me immediately, *Pappa*?" I ask – only slightly upset, for my joy is too great.

"I was under the impression *Mamma* told you, *my kind*," *Pappa* apologises as well. Together, we praise the Lord. My cup overflows.

Pappa hands the phone to *Mamma*. We discuss the results excitedly, and *Mamma* also apologises for not sharing the news with me sooner.

While I am still talking to *Mamma*, I am reminded of our last day on Orewa Beach and the storm which blew our tent over the caravan's roof. The only part of my healing board that blew away was the lyrics to "I raise a hallelujah" and the MRI results with phase four cancer all over it.

"What was the date of the last scan, *Mamma*?" I ask.

Mamma scrolls through her diary. "*My kind*, it was 19 October."

I draw in a sharp breath, my eyes pop, and I look up. *Lord!*

The storm, which blew the MRI results and the lyrics to "I raise a hallelujah" away, took place on 16 October!

You told me in advance!

Suddenly, I do not mind that both my parents and my sisters forgot to share the news with me – for I knew it first! The Holy Spirit Himself revealed it to me in a storm.

Maybe God wanted to test me to see if I would trust Him while the storm was at its worse, with a death sentence hanging over Pappa's head. Maybe He wanted to see if I would trust Him enough to step out of the boat in the middle of the storm and risk sinking. Maybe He wanted to see if I would keep my eyes focused on Him amidst the threatening waves and would ask Him to help us, simply because I knew He could.

Not only did I end up safe and dry in the arms of Jesus, but He also confirmed that the storm blew the scan results out of my healing board that day because He is indeed the God of miracles.

To top it all – the Springboks won the World Cup today!

Chapter 11
Jou Rooilippies, Jou Suikerbek

Mamma: "Lij, please remember to write the invite for our seventieth?"

I roll my eyes. Like so many times before, *Mamma* waits patiently (but constantly!) for my creative juices to kick in. *Mamma* always asks me to write invites for parties weeks in advance and then continues to remind me every time we speak.

Inspiration is not available 'on demand' as any writer or artist knows. Sometimes, the inspiration arrives at midnight; sometimes, whilst I'm driving. Some of my best writings were done just outside the print margins of our church newsletter, in my car, in the parking lot, just before I got out for my daily workout at the gym. Other times, the waves delivered the words during my walk on the beach.

One such invite was for their Silver wedding anniversary. It was for a surprise party at my auntie-sister-friend's house in Pretoria. Another one was for their thirtieth wedding anniversary. This one was a surprise party at other sis' house. Over the years, there were countless kitchen teas, baby showers, baptisms, and birthday parties.

Today, 25 October, the words arrive on the sunbeams pouring through my caravan bedside window. I scribble a bit, think again, rub it out, and scribble again until I'm satisfied. Excited (and relieved!) I send the funny, silly invite to *Mamma*:

Chris and Petro are having a BIG party this year!
For 70 precious years, they've been spared!
Come join us for a joyous feast,
And enjoy a joyful togetherness feast!

Saturday, 23 November is the BIG day,
To celebrate their birthdays and laugh away.
Come! *Kuier lekker*! We'll see you at ten.

At President Kruger Street Nineteen A.

RSVP before 15 November: *Tannie* Piet

Mamma: "*Dankie, my kind. Ai*, I cannot wait until next month. Oh, my! We are already talking about *next* month!"

*

The following two weeks are hurried and *deurmekaar*: We prepare for a seven-week trip to South Africa; we are notified that we need to take occupation of our new house two days before Niki and I fly out; we try to finish our homeschool goals for the year; write tests and … no, please don't ask! I have no idea how I am able to tick off each day's boxes, but, miraculously, the plane jerks as its wheels touch down at Jan Smuts[132] Airport … (better known these days as Oliver Tambo International.) It is 17 November 2019, and neither *Pappa*, nor the rest of my South African family suspects that I have already arrived!

Just before Niki and I boarded the plane, I sent Lin a colourful arrival notice via WhatsApp®:

ARRIVALS QF63.

We agreed that other sis also needed to be surprised. Lin is supposed to show other sis the notification just before they take the airport exit. Other sis was under the impression they were going to pick up something for *Pappa* and *Mamma*'s party. Only once she saw the WhatsApp® notification did she start wondering whether they were going to collect me from the airport. LOL!

*

During my childhood, whenever *Pappa* and I would go for a walk, he would take my tiny little hand into his, and it would disappear inside his big, warm grip. Nothing in the world could make me feel safer. With a lively twinkle in his eye, he would look down at me and start singing: *"Jou Rooilippies, jou Suikerbek! Waarom is ek na jou so gek?"*

Throughout the long flight, I prayed continuously: "Lord, please let me be in time to embrace him once more, to sing our song once more and to put my hand into his big, warm hand once more!"

Excitement builds up while Niki and I are waiting for our bags. Finally, we start walking down the familiar corridor. There, in front of us, the big glass doors are waiting. Niki calls out and asks me to wait, but I can't – not anymore! The automated glass doors slide open. There they are! I leave the trolley right in front of the exit, Lin ignores the line on the floor, and we fall into each other's arms!

The joy of seeing my sisters again is … indescribable! We laugh and cry simultaneously. They are amazed at how much Niki has grown since they last saw her. In between all the chatter and catching up, we pay for the parking, find the car, and load our bags. Before long, the airport is only a reflection in the rearview mirror.

The drive to Middelburg feels endless, despite our non-stop talking. Without making a sound, I continue to murmur: "Lord, please let me be in time!"

Finally, we pull into the large, wide driveway. Quietly, we slip into the house through Lin's front door. *Mamma* is already there to welcome us. Our tears flow freely as we cling to one another. Whispering, we walk through the shared laundry to *Mamma* and *Pappa*'s side of the house. I am unable to wait any longer. I take a deep breath as I enter the kitchen and start to sing: "*Jou Rooilippies, jou Suikerbek!*"

Suddenly, after months of the enemy trying to convince me that I won't see my dad again in this life, I am in time! In time to walk into the lounge, while I continue to sing: "*Waarom is ek na jou so gek*?" *Here, ek is betyds! Betyds om my pappa weer vas te hou!*

Pappa jumps out of his big, maroon recliner. I fall into his arms, and we cling to each other as though our lives depend on it. No words... *Pappa*'s and my own shoulder are getting soaked by each other's tears. My heart shouts repeatedly: *Dankie, Here! Thank you, Lord! Dankie, Here!* Then *Pappa*'s voice sounds softly in my ear: "*Dankie, Here*!"

"For where two or three gather in My Name, there am I with them!"[133]

*

18 November: The nephew, whom I loved first, turns twenty-one! I cannot remember when I last attended one of his birthday parties. What a blessing to be here tonight!

A few weeks ago, I sent him an SMS:

> "A little birdie told me that a very attractive young man will be turning twenty-one soon. Is there any chance that you are able to tell me what he would like as a birthday present?"
> Mike: "Actually, I only want some love and attention," he answered, partly embarrassed, partly jokingly.

My creative juices kicked in, and I asked Lin to buy me an extra-large gift box, since I wouldn't have had time to do so between my arrival and his birthday. I popped into a $-store before I left New Zealand and bought a few hundred multi-coloured craft hearts. On each one of them, I wrote either "love", or "attention". I also purchased a few party poppers and other kid's birthday accessories, together with a very special twenty first birthday card.

Back to today. I ask Lin if she would drive me to the nearest ATM, where I withdraw some cash. *Mamma* had mentioned that Mike would like to buy himself a car and that everyone had decided on financial contributions to his 'great cause'. As soon as we return to my parents' house, I put the money inside the birthday card and hide the card at the bottom of the gift box. I fill it up with hearts and birthday accessories full of love and attention.

Before I can rub the jet lag from my eyes, the evening is upon us, and we are on our way to Mike's birthday dinner. For the first time in twelve years, I can attend a family member's birthday party, other than my immediate New Zealand family's. We laugh and catch up, make toasts, and take several pictures. I sit between my *pa* and *ma*, with a protective hand on each one of their laps – like a mother hen. *Dankie, Here!*

*

Early morning, 20 November. I wait impatiently for Lin and her kids to come over, so we can wake *Pappa* with a birthday song like we used to do when we were kids. The moment is overwhelming. After the singing and well-wishes, I withdraw to my room, overcome by emotion.

The big party to celebrate *Pappa* and *Mamma*'s seventieth birthdays has been planned for Saturday. Accordingly, we decide to share a simple meal together and enjoy some cake this evening – just my parents, my sisters and their families, Niki and me. Just like I used to do as a little girl back in Tzaneen, I am already waiting to welcome Pappa home when he walks into the door earlier than usual. (After all, it is his birthday and the one whom he loved first is waiting for him!)

Without saying hello, he announced with beaming eyes: "I received some good news today. The blood test results arrived today". (A few days before I arrived, *Pappa* had gone for blood tests.)

I start to jump up and down like a little girl.

"My PSA-count is down to zero point six!" From thirty-six to zero point six!

We celebrate joyfully! Only my sisters, their families, *Pappa* and *Mamma,* Niki and I, to celebrate the festival of joy, victory and grace: scientific, medical proof that my dad is cancer free! We eat cake and make several toasts; we laugh and cry and sing and dance. It is a holy, heavenly festival of joy!

I lock eyes with *Pappa* and put my hand over his big hand: "*Pappa*, today the kingdom of heaven became tangible among us!"

Pappa looks at me with eyes full of gratitude and says: "I know exactly when I was healed."

Everyone listens with anticipation. "I cannot tell you the exact date and time, but a man walked into my office a few weeks ago. He too, battled prostate cancer. We started talking, and he was not very hopeful. I locked eyes with him and said: 'I know the plans the Lord has for me! It includes a future and expectation[134], and God's Word cannot lie!' The moment I said those words out loud, the pain left my body – and it didn't return!"

LORD! Lord! When You saw the faith of Chris Viljoen's daughters, You told Pappa: "Get up and walk! Be healed!" But it was only when Pappa SPOKE Your life-giving, healing words that he felt the illness leave his body.

Whenever we win our war, we should never neglect our feast of joy. Like that one leper, we need to turn around and say: "Thank you, Lord!" Like the Passover Festival when the Israelites celebrated their exit from Egypt and the end of their slavery; like the festivals of harvests at the beginning and end of the harvesting seasons; like the Purim festival to celebrate being saved from total annihilation, we need to celebrate our festivals of joy to give thanks for every bit of grace at the end of a devastating storm.

A festival of thanksgiving not only provides us with an opportunity to be reminded of God's faithfulness, but also to relax after a strenuous journey and to have a *lekker* party!

For a few days, we ride the wave of healing victory. Our joy knows no end. But little foxes were already busy ruining the vineyards. To tell you the truth, they started their shenanigans during our festival of joy.

Chapter 12
Ebenhaezer[135]

The next few days are filled with arrangements for the upcoming weekend. I often see *Pappa* and *Mamma* whispering rather worriedly. Maybe it is about the financial burden of the party so close to yearend? I offer to pay for the flowers and help *Mamma* wherever I can.

Together, we choose some table decorations at a local party hire shop. *Mamma* looks exhausted and overwhelmed. She is truly thankful for *Pappa*'s healing, but she neglected herself. I stare with concern at the dark circles underneath her eyes and the tiredness behind her shining blue irises.

"*Mamma*, shouldn't you go for some blood tests? Have you had your sugar levels tested?" She blows my concern off like she always does. When I insist upon an answer, she answers in a pretend drawling voice: "*Ja-ja-nee, goed, my skat.*"

This saying is from one of the plays we went to watch with my parents when I was still at school. The main character in the play was often scolded by his wife. With a submissive, drawling voice he would answer: "*Ja-ja-nee, goed, my skat.*" It became a family saying whenever one of us would nag a family member to do something, and the latter pretended to agree in order to keep the peace. Whenever *Mamma* answered me with this saying, I knew any attempt to try and convince her, would be futile.

Finally – or rather – sooner than we wanted it to – Saturday is upon us. Yesterday a large marquis was erected in the back garden and tables and chairs were dropped off. Early morning, we started moving the tables and chairs according to *Mamma*'s instructions. I cover the tables with the rented tablecloths and start to lay the tables. Many plates, many cutleries, and many glasses. Our step counters work overtime.

Between *Mamma* and Lin, they have more than enough glasses and plates, so no need to hire crockery. We load trays full of glasses and carry them down to the garden, one by one. Some of the fancy glasses Lin received as wedding gifts, are on the first tray I need to carry. Four large bowl-glasses with green coloured stems were a gift from one of her closest friends. As I pick up the tray, one of these glasses topple over and its stem breaks in two. *Please, let the earth swallow me whole!* I feel terrible. Lin replies with: "No, worries, they are only glasses." Mortified, I refuse to carry any more glasses: "I am too scared I will break more of them!"

With an angry frown between her brows, she scolds me: "*Man*, you did not receive 'a spirit of fear and timidity, but of power, love, and self-discipline'!"[136] She insists that I carry the rest of the glasses down to the marquis. Look, if you want to see my baby-sister in full-blown, angry teacher mode, you only need to utter the words: "I am scared!" I take a deep breath, blow it out slowly and say a quick prayer. To my relief, I break nothing else.

*

Before long the first family members arrive. The buzzing of voices grows louder as more cars arrive. And I am here! Just two months ago *Pappa* was too ill to fly, unable to sit down without experiencing severe pain and my two-year long dream to fly them out to New Zealand for their seventieth birthdays went up in smoke. It felt as if the storm was destroying us.

Nevertheless, we never took our eyes off Jesus. I might have shut my eyes tightly at times like I did in the *spooktrein*, but I never let go of the seam of Jesus' garment. Whether I was crawled up on His lap or lying flat on the floor in front of His feet – I never let go of Him during my storm.

That was our saving grace – neither the cancer bush, nor the healthy lifestyle, nor hormone therapy, nor bio-plasma scans could heal my dad. Yes, all these factors contributed to strengthen his body and to destroy the cancer cells, but *Pappa*'s healing was from God. The doctors couldn't believe it, but we could!

The whole of Saturday we celebrated an indescribable festival of thanksgiving. Family members whom I haven't seen in ages embrace me. We brag about our kids and laugh about our wrinkles. *Lord, You blessed us abundantly!*

Before we know it, it's time to eat. Slowly but surely, we ask everyone to take their seats. Champagne corks start popping and flutes are being filled. Table placings are such that my siblings and I sit right across from *Pappa* and *Mamma* in the middle of the long table, with our families next to us. The rest can sit wherever they want to.

As I am getting up to propose a toast to *Pappa* and *Mamma*, *Pappa* hushes me and stands up. He wishes to welcome everyone himself and to deliver his testimony. Obediently, I take my seat again.

Look, if there is one thing the Viljoens love, it is talking – especially to be the one talking. *Pappa* takes the lead when it comes to speeches and toasts. Needless to say, everyone wipes away a tear. The champagne is begging, but the guests are in for a surprise if they think they will be allowed to start sipping champagne anytime soon. I hope they are sitting *lekker*!

"Dearest Family and Friends," I begin. "I have to make up for eleven years of missing out on birthdays and special occasions. So please prepare yourselves for a long, emotional speech. If you are thirsty, grab a drink; if you need to go powder your nose, do so quickly; if you are standing, you may want to take a seat.

"I have much to say – not only because I am a Viljoen, but also because the two people whose seventieth birthdays we are celebrating today – my *pa* and *ma* – are who they are. For this reason, I decided to write down what I need to say. I'll try my best to stick to my notes, otherwise we are going to meander down many a pleasurable footpath in memory lane and you may never get to eat today."

A few family members have a chuckle and make themselves more comfortable.

"Because my *pa* and *ma* are who they are, the enemy tried his best to prevent us from gathering here today." I decide to bring some comic relief to my summary of our year of crisis by relaying it in the form of a silly poem:

"Friday, April twelve, a steel gate fell on *Pappa*'s head.
Out of nowhere seven angels appeared to lessen the threat.
As the devil's ice-cold fist clenched my heart,
A stronger voice hushed: 'Be still! And know - I am God!'
Warrior angels lined up for war,
And before most could visit, *Pa* was out the hospital door.
Within weeks bad news tried to rob our joy.
But covered by Jesus' blood,
The illness had to flee on foot!"

I take a deep breath and look at the family members around me.

"Today, we have a Guest of Honor. Not only is He our very best Friend, but on Wednesday, 20 November – on *Pappa*'s birthday – He delivered a gift with ribbons reaching the heavens! When *Pappa* opened that gift, the Kingdom of Jesus Christ, our Lord, became tangible among us! The latest medical report showed a PSA-count of zero point six. My dad has been healed from cancer and the last scan was clean!

"Our family celebrated on Wednesday! We sobbed as tears of joy ran down our cheeks; we danced and rejoiced and gave thanks repeatedly.

"Not too long ago, Pappa's PSA-count was thirty-six and a scan showed that the cancer spread though out his body. We literally drew red lines over that death sentence and chose to stand on God's promises. We reminded Him of all Jesus' miracles, and we believed that He *wanted* to heal *Pappa*.

"In the presence of the sceptics *Pappa* quoted Jeremiah 29 verse 11[137] and said: 'I know the plans the Lord has for me: plans to prosper me and not to harm me. He declared that He has a hope and a future for me, and God cannot lie!' That very moment, the illness left my *pa*'s body.

"Jesus himself said: 'For where two or three are gathered together in My Name, there am I in the midst of them' and 'the heartfelt and persistent prayer of a believer can accomplish much [when put into action and made effective by God – it is dynamic and can have tremendous power]!'[138].

"We stood together and carried each other through this. When one of us fell flat on his or her face, the others carried us.

"We raised a hallelujah in the presence of our enemy.
We raised a hallelujah louder than the unbelief.
We raised a hallelujah and death lost its hold on us.
We raised a hallelujah and heaven came to fight for us![139]

"You are gathered here today to celebrate so much more than *Pappa* and *Mamma*'s seventieth birthdays. Today we are celebrating that the Kingdom of Heaven is among us!

"2019 did not turn out the way I planned. I had to forget about my well thought out plans and I had to learn to 'let go and let God'. Alas! Here I am today to celebrate this triple-feast with you all!

"My journey did not include any smooth sailings. The past two months were quite an adventure!"

I decide it is time for a bit more comic relief:

"Our house in Hamilton - suddenly sold.
The purchasers wanted to move in –
we had to pack up and go!
Short-term, furnished homes were hard to come by.
Accommodation for 11 Alpacas and two cats – you give it a try!
So, we camped in our caravan on Orewa beach.
Homeschooling and house hunting; solutions out of reach!
The first three days camping seemed smart,
But four weeks later, we were blown apart!
We catch flying poles – preventing a dent,

While trying to rescue our upside-down tent!
Thus was the end of our camping event.
Grace upon grace – we found a place to rent.
We jump through a hoop – and the sale goes through.
No longer homeless – we found a real house!
We receive a settlement date two weeks before we fly:
Two days before I fly, I need to occupy!
At my wits end, I send *Ma* a text:
'Too little time! I'm trying my best!'
Four little words is all that I need,
To take a deep breath without losing speed:
'*My kind, jy kan*!'

"And behold, here I am! All the furniture arranged! Boxes are all over the place, but we have moved into our Auckland house and arrived in our country of birth just in time to celebrate today with all of you and especially with the two VIP's of the day: *Pappa* and *Mamma*."

I take another deep breath and shift my focus to *Mamma*, here, right in front of me. Her loving, shining blue eyes lock with mine and I am filled with the courage to start without tears:

"*Mamma* ...

"Thank you for being my first safe haven. With roughly two hundred thousand abortions a day[140] a mother's womb is the most dangerous place to be nowadays. Thank you that yours was my safe place.

When I look around me, I often see people who most likely never prayed before. Thank you for teaching me how to pray, *Mamma*.

"Ever since I can remember, I loved reading. I thought people who didn't like it, were rather clueless, for you can escape anywhere in the world by opening a book! One day I asked one of the non-readers: 'Didn't your *mamma* read stories to you when you were little?' I meant it as a rhetorical question but was shocked to receive the answer: 'No, she didn't.'

"Over the years the realisation dawned on me: some *mammas* just don't do it! Thank you, *Mamma,* for reading to us daily – until the day I indignantly told you: 'I can read for myself!'" *Mamma*'s eyes are smiling, but with a look that only the mother of a very stubborn, wilful child can understand.

"One takes so many things for granted. I am homeschooling Niki," I add while looking at my extended family, for I am not sure all of them are aware of the fact. "Whenever somebody would fail to do something properly, lack certain knowledge, or act inappropriately, she would comment: 'I guess their *mammas* didn't homeschool them.'

"Over the years I realised: Some mothers simply didn't teach their children a lot of the things mine taught me! Some people may not have had mothers who could teach them; some mothers probably didn't know any better themselves."

I lock eyes with *Mamma* again and noticed that her gaze didn't leave my face once, but her shining blue eyes have a few more sparkles than before.

"That is why I want to say today: Thank you! Thank you, *Mamma*, that you taught me how to make my own bed, how to do the dishes, how to bake and cook, about gardening and spring cleaning and the unwritten rule: "'*'n Mens doen dit nie*.'"

I look at my extended family and testify with conviction: "My *ma* taught by example that one does not hang around in pyjamas, nor wear your slippers all day, nor do you leave the house without looking presentable."

I return my gaze to *Mamma* and my memories go far back: "Thank you for all the clothes you made for me. When there wasn't money to buy new clothes, *Mamma* never failed to ensure that we were properly dressed. Thank you for the special birthday parties – whether finances allowed it, or not."

With pride I tell them about the *Mamma* of my childhood: "My *ma* always made the party decorations herself. There was always a special theme of my choosing and she always did it with so much love and care."

I look into *Mamma*'s eyes: "Thank you, *Mamma*, for being there when we were ill and for all the Marie-biscuits and rooibos-tea to medicate and heal. It had much more healing power than antibiotics," I wink.

"Thank you for driving me around – to school; to ballet classes; to athletic practice; to orating and debating events and to drama repetitions. Thank you for being there when I hit rock bottom. Like the time when I was so ill during my bar exam. You struggled with me through every day of that battle." *Mamma*'s eyes reflect the memory – the heartache and the pieces she so desperately tried to pick up and hold together.

"Guys, I was so ill!" I turn my focus back to the family audience. "I studied until I couldn't see straight anymore. Then I would tell *Mamma*: 'I am going to have a nap now. Please wake me up just before we need to leave for the next exam.' When *Mamma* woke me, I would get up, gulp down a Red Bull and by the grace of God and *Mamma*'s four words: '*My kind, jy kan*!' I would tackle the bar exam."

Tears of remembrance gather in my eyes when I look at *Mamma* and say: "I remember when you picked me up after that last paper. When I got into the car, I looked at you and said: 'We did it, *Mamma*!' *We* … not I alone …" I swallow a few times and pretend to consult my notes. *Mamma remembers …*

"Thank you, *Mamma,* for having the courage to bear two more children after this stubborn, wilful one made her appearance, otherwise I would have missed out on all the adventures I shared with these two sisters. Above all: thank you for staying with my dad throughout all of life's challenges."

I look at all the women around me – women who know the cost – whether they decided to stay or whether they decided to leave. Everyone is carrying her own load, has her own story and her own glory. Just like my *ma* and I.

"*Mamma* is the prime example of a woman who supports her husband in everything he does, even if she does not agree with his choices. Believe me, children are often more aware of their parents' challenges than the parents realise." My eyes assure *Pappa* and *Mamma* that I understand the sacrifices each of them had to make over the years.

"I have been a witness to this marriage for forty-six years. Most of you are aware of the fact that I had to do a few trial rounds, before I could conquer this thing they call marriage. However, one of the main reasons hubby and I have been married for almost fifteen years now, is the example of *my ma* and *pa*."

A few family members have a chuckle at this remark. Their eyes, however, are filled with love and compassion.

"*Mamma* and I made a few trips down memory lane this week when we perused a box full of old photos. I discovered a poem I have written on Mother's Day 2002 among the photos:

Simply my *Ma*

Safe and sound, hidden inside
your womb-warm blanket.
You have loved me first –
I was the first you carried underneath your heart.

The first to climb onto your lap,
and see mummy-tears glistening in your eyes.
With childlike innocence I bombarded you
With some impossible grown-up questions.

Simply my *ma* –
and I – your very first child!
In you – through you – from you.
Flesh from flesh – forever bound together!

Look at me – look inside me:
You have loved me first!
I was the first to break your heart.
You begged for mercy on my behalf – first…

And now I look at you – and I see you
As the most beautiful woman to ever trod this earth!
You were given the most difficult blood to love.
Your blood – each with a mind of its own.

And our other blood – our pa –
Would ask even more than us.
But you – our *ma* – ever so humble
Gently carried every one of us.

Here, I stand in front of you – now a woman, just like you.
I look at you – most beautiful one –
And with child-like angst I ask my *ma*:
Will I ever be able to carry my own with such grace?

But then my eyes meet yours;
your gaze penetrates my soul
and I see the most beautiful part of me!
This *ma* – mý *ma* – have loved me first indeed!

With faith, hope and love I come to you today,
And wish you the most blessed seventieth birthday!"

Mamma can barely contain her tears. I pray I have said everything she needed to hear, everything she deserves to hear. I do not want to sing her praises at her funeral. I want to look into her eyes and tell her right here and now everything she needs to hear!

Another deep breath gives me the much-needed courage to continue. My eyes move to the man I have loved first – and never stopped loving.

"*Pappa*…

"A *mamma* may very well be her child's first safe haven, but a *Pappa* is the first man his daughter will ever love – and for most of us, he remains her first and greatest love throughout her life.

"My earliest memory of *Pappa* is when I waited for him at the bottom of the staircase each day. That is, after I continued to nag throughout the morning: '*Mamma*, when can I go and wait for *Pappa* by the stairs?'

"My earliest memory of wintertime, is of *Pappa* in his brown, sheepskin jacket, with my small, frozen little hand tucked safely into his big, warm hand. It was the safest place in the world!"

Jokingly, I address the others: "At school I discovered most kids found history boring. What a shocker!" The shoe fits some of my audience and they pretend to be outraged. "I, on the other hand, had a dad who would put me on his lap and told the heartbreaking story of Little Rachel de Beer so beautifully, that it would remain my favourite, despite all the tears it caused. Within me I carry a treasure chest filled with the most beautiful stories – many of them told to me by my dad.

"I have visited most of the places worth visiting in this country. My *pa* believed that whenever we went on holiday, the road would take us on a journey – and not merely to a destination.

"My love for song and music is also thanks to *Pappa*. Viljoen-family *kuiers* are always known for musical instruments and *lekker* sing-alongs. 'Sweet Lijlanie' will always be my favourite song and the one my *pa* plays and sings best. A close second is 'Sweet Caroline' – our favourite dance song.

"A few of you share one of the other great loves *Pappa* installed in me: motorbike riding. I was just a little girl when *Pappa* started to take me on motorbike trips with him. My poor *ma* – because no matter how special the bike rides with my *pa* was, I would never allow my little girl on the back of a motorbike! Fortunately, *her* dad hasn't bought one yet!" I add with relief.

Pappa's eyes reflect my light blond hair waving in the wind from underneath the helmet. In our minds we visit many a ride down memory lane. "Thank you for all the special road trips where it was only you and me – with no words necessary." *Pappa* swallows hard on the lump in his throat. I look at the others and share another funny story.

"My *pa* also taught me how to bunk school!" A few shocked, unbelieving faces stare back at me, for everyone knows how strict my *pa* was.

"My love for long distance road trips started at an early age – when most other kids despised them. *Mamma* would wake me up while it was still cold and dark outside. She told me to dress in shorts, a t-shirt, and a tracksuit. That way I could take off some layers when it got too hot in the car. Like a VIP I would sit on the front seat of the *bakkie* next to *Pa.* With the cargo tray loaded high with high jump mats, we would hit the road to one or the other countryside high school. The journey always included a few highlights:

"The first highlight was when we stopped at a *padkafee* to buy some cold drinks (and to visit a much-needed loo!). The second highlight was when the *tannie* at the school office served us tea while I am sitting all grown-up next to *Pappa* in the principal's office. But the best of the *kafee*-stop, was Pinball! Believe me, if the *kafee* didn't have a pinball machine, we had to go find another, since pinball was a vital tradition on these road trips.

"It was this road trip tradition that caused my dad to teach me to bunk school. One morning I was sitting at the breakfast table, all dressed up in my school uniform, ready to go to school, when *Pappa* mentioned that he had to go and drop off some high jump mats again. What a shocker when I was told that I am not allowed to go with him anymore, since I apparently have to go to school now. I think *Pappa* could see how traumatised I was, for he gave me one look and instructed me to get out of my school uniform and into something more comfortable. I was going to bunk school – officially! Officially – for he gave my school principal a call and informed him that I was going to bunk school in order to deliver some high jump mats and to play pinball with my *pa*!"

I can see that not everyone knew this story and that some are even doubting its truthfulness. But those who remembered, had a naughty twinkle in their eyes.

"Everything *Pappa* taught me, he taught me well. Well, I guess it had to be well, for it stuck with me. (Except for the bunking – because my *pa* decided to become school principal himself. All the teachers among us know all too well: you cannot trick a teacher with his own tricks!)" Everyone has a good laugh – especially the teachers!

"*Pa* taught me that when tragedy strikes, there is nothing wrong with a good cry. But then, you need to wash your face, make yourself presentable, straighten your shoulders and look the world square in the face. Obviously, all of this needs to happen with your hand tucked firmly in our Heavenly Father's Hand.

"That is why we can conclude this year with straightened shoulders, an expectation and hope for the future. A Viljoen does not hide in the boat while the storm is raging. We focus our eyes on Jesus – and walk on water! Okay, I'll admit – we are not very good at walking – we run most of the time!"

My family responds with laughter, for they know all too well that few people can keep up with what my *pa*, Lin and I call our 'normal walking pace'.

"We may end up soaking wet on the other side, but we always conquer!

"Believe it or not, both *Pappa* and *Mamma* shared quite a few wise sayings during the course of their lives," I tease. "Even though there may have been some rolling teenage eyes behind their backs, or a frustrated '*Ag, Ma*!' I have learned to return to their words of wisdom ever so often over the years.

"I am forty-six years old, and I still have both my parents, who are still married to each other. Guess what? They don't look too unhappy!" I tease again, while my eyes are laughing with *Pappa* and *Mamma*.

I look at my family more seriously: "Not too many people my age can still say that.

"Lift your glasses and toast with me. But wait! Before you get too excited and start drinking too soon – I am not done!

"Lift your glasses with my sisters, our hubbies and me, and salute our *Pa* and *Ma*;

"Lift your glasses with Rudo and Niki, Michael, Nathan, Gadduell and Keara, Christiaan and Linieke, and salute their *Oupa* and *Ouma*;

"Lift your glasses and salute *Chris* and *Petro* – a brother, a sister; an uncle, an aunt; a godfather, a godmother and a friend.

"*Pappa* and *Mamma*, congratulations on seventy full years. May God, our loving and compassionate Father, bless you abundantly with HEALTH, a long and happy life – and many great grandchildren!

"To my *pa* and *ma*!"

Epilogue
The Desert

Little foxes…

Many people chose to respond in praise and worship when hearing Pappa's good news. On the other hand, there were those who were supposed to believe in miracles who chose not to believe. Some of them were rather close to my dad. Some even looked at *Pappa* as though he was in denial; others said, "If he really had cancer …"

Fair enough – the medical specialists could have erred. I guess the blood tests showing a continuous rise in PSA-count were faulty time after time. The specialist who did the biopsy and observed how the cancer had spread into the surrounding glands and bone probably needed new glasses. The MRI scanner probably had dirty lenses (explaining away all the black spots) and the interpreter of that MRI-scan was most likely just another end-time prophet of doom! Oh, not to mention the oncologist who told Pappa that chemo was his only hope; the medical aid who didn't want to waste money on radiation for a person who didn't have long to live anyway … and all the symptoms, pain and discomfort was probably just a figment of my *pa*'s imagination? Humph!

One evening, a mere week after Pappa was declared cancer-free, he mentioned to me in an irritated voice: "People don't believe that God has healed me." He was angry and felt that they were dismissing God's miracle. It was as if some of them preferred pity parties to thanksgiving festivals.

Pappa's initial joy became tainted with upset and anger. One morning during my quiet time, the Holy Spirit guided me through the gospel of Mark. He reminded me that Jesus told His disciples no less than three times during His three-year ministry that He was going to be killed but that He would rise on the third day:

> Mark 8:31: "He then began to teach them that the Son of Man must suffer many things and be rejected by the elders, the chief priests and

the teachers of the law, and that He must be killed and after three days rise again."

Mark 9:31: "He was teaching His disciples. He said to them, 'The Son of Man is going to be delivered into the hands of men. They will kill Him, and after three days He will rise.'"

Mark 10:33-34: "'We are going up to Jerusalem,' He said, 'and the Son of Man will be delivered over to the chief priests and the teachers of the law. They will condemn Him to death and will hand Him over to the Gentiles, who will mock Him and spit on Him, flog Him and kill Him. Three days later He will rise.'"[141]

The Holy Spirit smiled at me with compassion and said: "Do not be angry at those who refuse to believe. My disciples, whom I have chosen by hand, learned from Me for three years whilst sitting at My feet, and experienced the miracles I did in person. *Three* times I told them clearly what would happen, but the moment I died on that cross, they forgot everything I taught them!

"When Mary told them early that morning on the third day: 'Jesus has risen!' they refused to believe her and did not remember what I had personally told them. Therefore, I appeared to them in person, where they were hiding behind locked doors. Poor Thomas refused to believe even the eleven – let alone a woman! So, I appeared to Him too and invited him to touch Me and to feel for himself that it truly was Me. If they, who lived so closely to Me, did not believe at first, have patience with those who struggle to believe. I meet each one where they are."

After *Pappa* returned from work, I shared this revelation with him. But he wandered off to the desert, isolated himself in his cave and started to struggle through deep matters of the soul. I desperately wanted our festival of thanksgiving to last longer!

Lovingly, the Holy Spirit took me by the hand and reminded me of Elijah, who also fled to the desert. After his majestic victory over the prophets of Baal, Elijah withdrew to his cave. With renewed insight, I studied Elijah's story in 1 Kings 18 and 19.

It explains how Elijah experienced miracle upon miracle. First, there was the fire from heaven that consumed his offering after he called upon the Lord once. That happened after the prophets called upon their gods for hours – to no avail! Then Elijah prayed for rain, and after three years of drought, it rained! Amid the pouring rain, God gave Elijah supernatural power to outrun the king's chariots! All of that happened in one single day.

The next moment, a *female* threatened to kill Elijah because he killed her Baal prophets, and he decided to flee to the desert. There he curled up underneath a bush and hoped to die! Eventually, he couldn't fight the exhaustion anymore and fell asleep. After he rested, an angel woke him twice and fed him. This food provided him with enough energy to walk for forty days and forty nights until he reached Horeb, the mountain of God. Three more glorious miracles that Elijah experienced... Yet, there on God's mountain, he was hiding in a cave.

It was as if Elijah was trapped in a spiritual dimension of miracle upon miracle. Miracles – indescribably huge! But the human Elijah fled to the desert and walked for forty days until he reached his cave. There God met with him in person, in all His Godliness!

Slowly but surely, I realised: This was my *pa*'s journey. I couldn't help him anymore ... He withdrew from me more and more, into the desert – into his cave – and I was unable to prevent it...

So, I came back home.

After our huge miracle, I also had a 'So what now?' moment – just like my *pa* and just like Elijah. Back home, I had to struggle through my own desert ... and the Lord came to fetch me from where I was hiding in my cave and brought me here, today, to tell you my story.

Your reason for your desert may look very different from mine. You may experience a time in the desert because you didn't experience an answer to your prayers.

Or maybe your war is continuing way too long?

Have courage! Even the Israelites entered the Promised Land after forty years. God doesn't care about our earthly goals or destinations; He concerns Himself with our eternal destination. To Him, our earthly 'goals and destinations' are a mere part of our journey.

Dear Reader, I do not have the answers to your hurts and questions. But what I do know, is that when you spend enough time at the feet of Jesus, if you put your hand in His, trusting Him, and if you are willing to walk this road with Him, one of two things will happen: either you will receive your answers in the end, and the pain will subside, or it simply won't matter anymore. Because at the feet of Jesus, you will find peace.

> "In the presence of the Lord there is joy;
> In the presence of the Lord there is peace;
> In the presence of the Lord there is healing.
> Lord, I stand in Your presence with my life!"[142]

Glossary

Abba - Hebrew word for "Father".
Afrikaans is lekker – Name of Afrikaans concert held annually in Auckland, translated "Afrikaans is fun".
Ag – Uhh!
Ag, ek is daar, man! - But I am there, man!
Ag, Here – Oh, Lord.
Ag nee – Oh, no.
Agge nee, man! - Oh no, man!
Ai - Interjection meaning "ah" or "oh".
Ai, toggie! Oh, dear!
appel-lief-julle-tert - A pun on the dish "appelliefie-tert" (gooseberry tart). Directly translated, the phrase means "apple-love-you-tart".
Asseblief – please
Baie dankie, my kind. – Thank you very much, my child.
Baie geluk, julle! - Congratulations, guys!
bakkie – utility vehicle – NZ "ute"; American "pick-up".
boer – farmer
boeresport – Traditional South African games played at carnivals, church or town festivals.
bokveld toe – Expression meaning "gone".
Bring die hoop weer terug – Title of Afrikaans song by Jo Black, translated "Let hope return".
cuppa – Kiwi colloquialism meaning "cup of coffee".
Dankie, julle – Thank you, you all.
deurmekaar – confused
Die een wat Pappa eerste liefgehad het – The one you loved first.
Die Hel – Translated "The Hell". The isolated, remote settlement in the Gamkaskloof Valley, located in the Swartberg Mountain Range, was so named by a traveller who referred to the experience of descending down into the valley as "hell".
Ek het Pappa baie lief – A more intense form of "I love you very much, Dad."
Ek is baie lief vir Pappa – I love you very much, Dad.
gedoente - Referring to "grotesque" headgear.
Getroud met rugby – "Married to rugby" - a popular South African soap opera.
Gibbor – Hebrew word which means mighty.
gogga – Commonly used to refer to insects or other creepy crawlies.
Ha! – Interjection showing contempt or disbelief.
Hallo, my liefling – Hello, my darling
hardekop – Metaphor meaning "stubborn" or literally "hardheaded".
hartkind – Child of my heart. A term of endearment for a non-biological child that one loves as if they were your own.
hartsvriendin – A close female friend who truly gets you – soulmate.
Here, ek is betyds! Betyds om my pa weer vas te hou! – Lord, I am in time! In time to embrace my father once again!
Hoor hier – Listen here!
Ja – Yes

Jehovah Rapha/Rophecha-- Hebrew Strong's 3068 and H7495 Yahweh Rapha means 'The LORD who heals' (see Exodus 15:26).
Ja-ja-nee, goed, my skat – "Ja-ja-nee" is similar to the Kiwi phrase "yeah-na", which more often means no, than yes. "goed, my skat" is translated: "Okay, my dear."
Ja, my kind – Yes, my child.
Jinne! – Interjection meaning "Gosh!"
Jou Rooilippies, Jou Suikerbek – An Afrikaans folk song. Directly translated it means "Little red lips, sugar-mouth".
Jou verlangende Pa – Your longing Dad.
Juffrou – female teacher
Julle! – Mates/Guys!
Jy't my gewys – You showed me.
kaalvoet – barefoot
kafee – roadside café or what Kiwis would call a dairy.
Kiewiet – Nickname, also the name of a bird.
Klingel, kletter, tingeling... (an onomatopoeic phrase describing glass shattering); *duisend splinters om my heen* – a thousand glass shards surrounding me. The sentence describes the dream of Dad's visit that broke into a thousand pieces, like a shattered glass vase.
Kruisgenerasie – Afrikaans Christian Church in Middelburg, Mpumulanga, South Africa. Translated "Crossgeneration".
lekker kuier - An enjoyable time of fellowship.
Lekker slaap, julle – Sleep well, you all.
'Lief julle almal baie! – 'Love you all very much!
Liefde – love
Liefling – Darling
Ma – Mum
maar, nou ja - but yeah
Mamma – Mum
Mamsie – Mumsy – term of endearment when addressing one's mother.
man – Translated "man" but used to show annoyance.
matriek – Final High School year.
my ander ek – my alter ego
my kind, jy kan – my child, you can
my pa – my dad
'n boer maak 'n plan – Afrikaans expression meaning "A farmer will always find a way to overcome any obstacle."
'n Mens doen dit nie! – Indicating inappropriate or unacceptable behaviour. Directly translated: "One does not do such a thing!"
Nee – No
Nie – not
Nou ja – Well, all right then.
O tog! – Oh dear!
Oe! – Oh!
Oom – Uncle. Afrikaners would not only address the brothers of their parents as "uncles", but also any male of an older generation, as a sign of respect.
Ooms – uncles.
Ouma – Grandma or *Nanna*

Oumagrootjie – great grandmother
Oupa – Grandpa or *Pops*
Oupagrootjie – great grandfather
Pa – Dad
padkafee – "Roadside café" or what Kiwis would call a "dairy".
Pappa – Dad
Poplap – A term of endearment, like calling a loved one 'babe'. Literally, ragdoll, from the idea that a child holds her ragdoll very dear to her and won't go anywhere without it.
rooibos – Popular South African herbal tea, translated "red-bush".
rumatiekbad – Natural hot spa, literally, rheumatic bath.
seën-sous – blessing sauce
Sewende Laan – Well-known Afrikaans soap opera, translated "Seventh Avenue".
Siestog – Sympathetic interjection, translated as "shame", but not in a bad sense.
Sjoe – Ouch!
Skepe – Ships
skoonma – mother-in-law
So ja – There you go.
spookasem – candyfloss
spooktrein – train of horror at the amusement park
Swaer – Used in a similar way to the Kiwi form 'mate'. Translation: "brother-in-law".
Tannie – Auntie
teen 'n slakkepas – Expression meaning "to go very slowly". Translation: "as slow as snails".
tokkelossie – An indigenous South African word meaning "ghost" or "evil spirit".
Transvalers – Residents of the pre-1994 Transvaal Province in South Africa.
veldskool – A school adventure camp full of obstacles and challenges that Grade 7 and 10 pupils used to attend before 1994. Translated: field school.
verdomde ding – stupid/useless thing
Vrede vind – Finding peace.
Vyf minute – five minutes
Waarom is ek na jou so gek? – Why I am so crazy about you?
Watwou wil – Expression meaning "How can a lifeless object have a will?" Thus, one should not say "it won't work", for it is human will that makes it work or fail, not its own will."
Wyk, satan – Be gone, satan!

Endnotes

---

[1] Josh. 1:9, HCSB.

[2] Rom. 8:26; 27; 34, NIV – own paraphrase.

[3] Jonathan David Helser, Melissa Helser, Jake Stevens and Molly Kate Skaggs. Copyright © Bethel Music Publishing.

[4] Jos. 1:9, NIV.

[5] John 10:10, NKJV – own paraphrase.

[6] Rikshaw – a small two-wheeled cart for one to two passengers, pulled by one person.

[7] Commonly used in South Africa to refer to swimwear.

[8] Ps.46:1, NIV.

[9] A Pace is a study unit of the Accelerated Christian Education (ACE) curriculum.

[10] Collapsible 1970's caravan with striped blue and white tent walls and side tents.

[11] Reference to a song by Carike Keuzenkamp, a well-known female South African singer of children's songs.

[12] Reference to a song by Laurika Rauch, a well-known female South African singer.

[13] Tabard is an insect repellent which comes in a roll-on stick or a lotion, used in South Africa and the UK.

[14] Oh, no!

[15] A young Afrikaans heroin, called "Racheltjie de Beer". (The '-tjie' at the end of the name is pronounced 'key' and is the diminutive form of the name.) Some say her story was only folklore, but many Afrikaners believe the tragedy actually happened.

(Check out https://en.wikipedia.org/wiki/Racheltjie_de_Beer)

[16] A South African chain store, similar to K-Mart.

[17] Nickname – derived from her Dutch maiden name, Graphorn.

[18] A general way to refer to Australasia.

[19] Kiwi slang, meaning something is easy. Kiwi's like to use incomplete comparisons, for example "tough as"; "good as"; "clever as"; "easy as", etc.

[20] Well-known Afrikaans poet (1873-1932).

[21] 1 John 4:4, The Message.

[22] A South African folk-rock ensemble formed in the 1960's.

[23] Short for Potchefstroom, a beautiful university town in the pre-94 Western Transvaal.

[24] 2 Chronicles 18:4, NIV.

[25] Is.41:10, NASB.

[26] Jes.49:16, NBG51 (Old Dutch Bible © Netherlands Bible Society 1951). Translation: "Behold, I have inscribed you on the palms of My hands" Is.49:16, NASB.

[27] A semi-desert valley between several mountain ranges in the Western Cape.

[28] Daleen Matthee, author of several novels with its setting in the Knysna Forest.

[29] H. Owens, Sweet Leilani lyrics © Chappell & Co., Inc.

[30] Matthew 21:22, AMPC.

[31] John 14:13-14, NKJV.
[32] John 14:27, GNT.
[33] NKJV.
[34] Matt. 8:2, NLT
[35] Matt. 8:3, NLT
[36] Heb. 4:16, NKJV – own paraphrase.
[37] The Devil Saw ME with My Head Down and Thought He'd Won Until I Said AMEN – A Prayer Journal Or Bible Study Notebook. Published 12 August 2019. www.google.co.nz/books
[38] Original author unknown. Published on Twitter by Fearless Warriors in Christ on 12 June 2018.
[39] Acts 3:16 & 4:10, NIV.
[40] Luke 10:19, NIV.
[41] Locke G. Sermon Quotes. Published 20 August 2019. www.sermonquotes.com
[42] De Edifier. www.deedifiertunes.com.ng
[43] Christian Today.
[44] Title of song written by Benny Andersson en Björn K Ulvaeus. Copyright © Universal Music Publishing Group.
[45] New Zealanders like to add "as" at the end of terms. "It's essentially a simile with no end." See https://ninaoutandabout.ca/new-zealand-slang/
[46] From *Andante, Andante* by Benny Andersson en Björn K Ulvaeus. Copyright © Universal Music Publishing Group.
[47] 1 Cor.2:2, NIV – own translation.
[48] Koki/cokey pen, felt-tipped marker.
[49] See endnote 3.
[50] Steven Furtick © O/B/O Amcos.
[51] Chris Tomlin, Ed Cash en Scott Cash © Capitol Christian Music Group, Music Services, Inc.
[52] See endnote 38.
[53] Eph. 6:12, KJV.
[54] Christian Today.
[55] Acts 4:20, Rom. 8:11, HCSB
[56] Matthew 8:2-3, HCSB.
[57] Luke 10:19, HCSB.
[58] Rom.8:27, HCSB – own translation.
[59] Steven Furtick © O/B/O Amcos.
[60] This is a move. Toney Brown, Tasha Cobbs Leonard, Brandon Lake en Nate Moore. © Bethel Worship Publishing.
[61] Ps.16:6, ESV.
[62] Mark 2:9-11, ESV.
[63] Prayerful Woman. Published on Facebook by Beauty from Ashes Ministries on 25 July 2019.

[64] v. 8-13, ESV.

[65] Ch.14 v. 32, ESV.

[66] Home remedy to ease a cold, made with aspirin, lemon, vitamin C, boiled water, and brandy.

[67] Prayerful Woman. Published on Facebook by Beauty from Ashes Ministries on 25 July 2019.

[68] NKJV.

[69] WEB.

[70] NET.

[71] The name of our housekeeper back in South Africa.

[72] HCSB.

[73] God willing.

[74] Hoffmann N. Sweatpants & Coffee: Affirmations for Anxiety Blobs (Like You and Me) 2021.

[75] "Big Move". The "Groot Trek" refers to a historical mass migration during 1833-1838 of the Dutch/French-Afrikaners from the Cape of Good Hope to the areas north of the Orange River.

[76] Word for You Today – Adaptation of The Word For Today BY Bob Debby Gas. Adapted by UCB UK. August, September, October 2019.

[77] Rom.8:26,34, NIV.

[78] Carman Licciardello © Concord Music Publishing LLC.

[79] Mark 11:22-26, HCSB.

[80] Taken from Song of Songs 2:15, ESV

[81] Brother-in-law.

[82] HCSB.

[83] Jo Engelbrecht and Johan Vorster. © O/B/O Capasso.

[84] This is a move. Toney Brown, Tasha Cobbs Leonard, Brandon Lake and Nate Moore. © Bethel Worship Publishing.

[85] Leaf C, Dr. *Switch On Your Brain.* 2015. Baker Publishing Group.

[86] ESV.

[87] v13, ESV

[88] Proverbs 16:9

[89] Joel Osteen

[90] Tierlantyntjies – translated.

[91] Rom. 5:1-5, NIV.

[92] Proverbs 19:21, NIV.

[93] Acts 3:16, HCSB.

[94] Joel Osteen

[95] See endnote 3.

[96] Derek Prince Ministries. Bible verses and Tags posted by Derek Prince. Accessed from www.biblenets.com

[97] Andrew Wommack

[98] 1 Cor. 14:32, ESV.

[99] Hebr. 11:1, ESV.
[100] Translated from "Jesus se Geloofs-Kosmosse. Dag 21." www.m.facebook.com/JesusseGeloofs Kosmosse/photos/dag-21
[101] Twizza Super School Series – a prestigious athletic competition, annually hosted in South Africa.
[102] "A thousand glass shards surrounding me." The dream of Dad's visit broke into a thousand pieces, like a shattered glass vase.
[103] v11-12, MSG.
[104] v10-21, HCSB
[105] v6, HCSB.
[106] 2 Cor. 10:3-5, MSG.
[107] Spiritual Warfare by Tonilee Adamson and Bobbye Brooks – www.biblestudytools.com
[108] From Eph. 6:16, HCSB
[109] Marvel Cinematic Universe Wiki. Captain America's Shield. Accessed from www.marvelcinematic univerce.fandom.com
[110] 1 Cor. 2:2, NIV.
[111] Joel Osteen.
[112] Original author unknown. Published on Twitter by Fearless Warriors in Christ, 12 June 2018.
[113] Eph. 6:12, KJV.
[114] Christian Today.
[115] Acts 3:16, 4:10 and Rom. 8:11 HCSB, personified.
[116] See Rom.8:11, NLT.
[117] Math.8:3, CSB.
[118] Luke 10:19, HCSB.
[119] Rom. 8:26, NLT.
[120] Chris Tomlin, Ed Cash en Scott Cash © Capitol Christian Music Group, Music Services, Inc.
[121] Jonathan David Helser, Melissa Helser, Jake Stevens, and Molly Kate Skaggs. Copyright © Bethel Music Publishing.
[122] Steven Furtick © O/B/O Amcos.
[123] This is a move. Toney Brown, Tasha Cobbs Leonard, Brandon Lake, and Nate Moore. © Bethel Worship Publishing.
[124] Prayerful Woman. Published on Facebook by Beauty from Ashes Ministries on 25 July 2019.
[125] Eph. 6:16-17, NIV.
[126] James 4:7, NIV.
[127] Jonathan David Helser, Melissa Helser, Jake Stevens, and Molly Kate Skaggs. Copyright © Bethel Music Publishing.
[128] Joyce Meyer Daily Devotional 6 May 2022.
[129] Mastercard's slogan during the late 1990's.
[130] Steven Furtick © O/B/O Amcos.

[131] A concrete tetrapod, which serves as a wave-dissipating harbour block.

[132] Oliver Tambo International Airport was called Jan Smuts International Airport prior to 1994.

[133] Math. 18:20, NIV.

[134] See Jer. 29:11.

[135] "Thus far, the Lord has helped us."

[136] Tim. 1:7, NLT.

[137] See NIV.

[138] Math. 18:20, KJV; James 5:16, Ampl.

[139] Jonathan David Helser, Melissa Helser, Jake Stevens, and Molly Kate Skaggs. Copyright © Bethel Music Publishing.

[140] www.who.int/news-room/fact-sheets/detail/abortion)

[141] NIV.

[142] Jodache Callan https://tabs.ultimate-guitar.com/tab/misc-praise-songs/in-the-presence-of-the-lord-chords-2693622